MCDOUALL STUART
HITCHES A RIDE

(UN)SETTLING ROAD TRIP
THROUGH THE CENTRE OF AUSTRALIA

ROSEMARY
CADDEN

Published by Dinkus Publishing
South Australia

Enquiries and book orders:
rosemarycadden@gmail.com

Front cover: the road features the modern Stuart of Bute tartan
Back cover: this portrait of John McDouall Stuart is digitised and kindly
supplied by the State Library of South Australia.
(SLSA: B501 – John McDouall Stuart, c.1860)

Book cover and interior design by
Arjan van Woensel and Marisa Garau

Illustrations by Jan Finlayson

First edition 2023
Reprinted 2024
ISBN: paperback 978-0-6459606-0-0
ISBN: e-book 978-0-6459606-1-7

 A catalogue record for this
book is available from the
National Library of Australia

DEDICATION

For Valmai Hankel, former Rare Books librarian at the State Library of SA, who would throw a swag in the back of her car and head into the outback—thank you for inspiring me to do the road trip, and for your encouragement to write this book.

And for everyone who has an adventure burbling inside.

I AM LEFT THINKING 'WHAT IF'

I pay my respects to the past, present and future traditional custodians and elders of this land. In particular, I thank the many Aboriginal Nations whose land I passed through on this road trip.

I acknowledge Aboriginal and Torres Strait Islander peoples as the land's first storytellers, and thank the many individuals who gave me advice on my efforts to acknowledge their Dreaming stories and to describe the land they've lived on for tens of thousands of years.

This book is my story, as a migrant, of learning about this country I now call home and, in particular, about our history since colonisation.

I am left thinking 'What if...'

What if the Colonising Commissioners who landed in South Australia in 1836 had brought with them the Letters Patent signed by King William IV – the founding document that recognised these Peoples were the rightful owners of the land?

What if, when we did learn about those papers, we had given them the full weight they deserved?

What if we had paid respect, as we do now, to those traditional owners and realised that, belonging as they do to the oldest continuing living culture on this planet, they had a thing or two to teach us about caring for the land?

TABLE OF CONTENTS

INTRODUCTION

As I looked up, up, up at the lanky statue of the diminutive explorer John Mcdouall Stuart in Alice Springs, my road trip to follow the tracks of this fellow Scot almost came unstuck.

The real journey, however, was just beginning.

McDouall Stuart, the first European to travel on horseback through the centre of Australia from Adelaide up to the Top End, was in his early twenties when he emigrated Down Under. So was I. He was born in Dysart, a fishing village over the Forth Bridge from Edinburgh, now absorbed into Kirkcaldy where I grew up, a gritty industrial town which, in my day, boasted a coal mine at each end and was famous for its linoleum factories. We were both looking for adventure, a new life, a new beginning. Our boat trips were 140 years apart. I've begun relationships on far flimsier grounds.

McDouall Stuart had stalked me for decades, from Adelaide to Scotland, from Scotland to Darwin, a commemorative cairn here, a newspaper article there. He was an underdog in the world of exploration and he lived in the shadow of Burke and Wills, a duo whose tragic deaths gave them a place in posterity. Today he is often confused for another explorer Charles Sturt. Those who *have* heard of McDouall Stuart are likely to focus on his drinking. I've been known to hold a wine glass in my hand, and so it came about that I chose to follow his route from Adelaide to Darwin to better understand this country I've called home for nearly five decades. After months of planning, preparation and dredging up outback skills I'd learned and forgotten, I set off on my month-long solo road trip.

1

It was a bumpy ride of pesky flies and strange sounds in the night, unexpected hazards and close calls. I marvelled at the resilience of the people I met along the way, who have made that stretch of land their home for a short or a long time; some connected since their ancestors arrived tens of thousands of years ago, others settling in more recent times. Even today, the mobile phone is a glorified alarm clock once you're a few hundred kilometres into the Centre from the coastline, and fresh water and electricity is not a given. There was little opportunity for self-reflection as I unpacked and repacked sleeping gear each night, shifted boxes of research material and tourist brochures I never got around to reading until much later, took thousands of photographs and taped interviews with locals.

The ride became bumpier still in Alice Springs, prodded as I was by that modern-day statue of McDouall Stuart, erected in Alice Springs at a time when calls to pull down colonial statues were reverberating around the world. Who was this man I was following?

The land, too, was not as I saw it. The 'bush' I once considered dry, dusty and frankly in need of a good tidy-up became a source of wonder. Are 'dull' and 'desolate' the words you conjure up with the words Arid Lands? I was that bunny.

Researching and writing this book was the bumpiest ride of all. It had me questioning myself, challenging my beliefs about settler colonisation and the environment, my biases and preconceived ideas. Research into McDouall Stuart's expeditions would expose the huge gaps in our knowledge of Australia's history—hidden, sidestepped, ignored. I grappled with trying to find 'the truth'.

Impressed as I was with the landscapes, up close and from afar, this is still no softly, softly travelogue.

While I followed the tracks of fellow Scot, John McDouall Stuart, and admired his tenacity and bush skills, I haven't put him on a pedestal.

This book, also, contains no answers. I started off this journey with the aim to better understand this country I call home. I hope you're like me and like to have more questions to ponder.

I constantly came across quotes that would push me forward. Here are but two:

The late Hilary Mantel, best-selling author of historical novels, once argued that 'facts are not truth (but) the record of what's left on the record'. I read this long after I grappled with how to deal with this nebulous beast called history.

'Chasing meaning is better for your health than trying to avoid discomfort,' is a comment I heard loud and clear on a TED talk by health psychologist Kelly McGonigal.

And just as I was putting a full stop to the manuscript that began back in 2015 (even earlier if I'm being honest), I was given the gift of a quote on the Book Club on ABC Radio National attributed to the late American author Cormac McCarthy, which I paraphrase here: 'Anything that doesn't take years of your life... hardly seems worth doing.'

PART I
GETTING

AWAY

From Adelaide
to Burra

STUART'S TRACKS

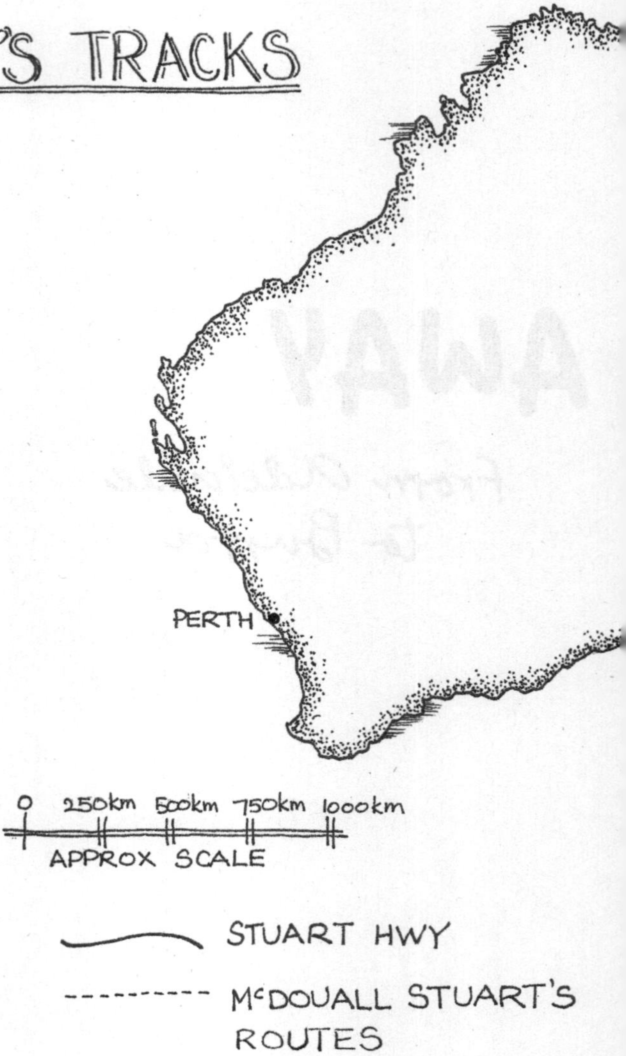

PERTH

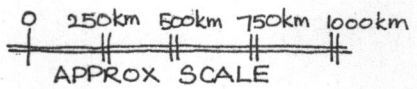

——————— STUART HWY

- - - - - - - - - - M^CDOUALL STUART'S
ROUTES

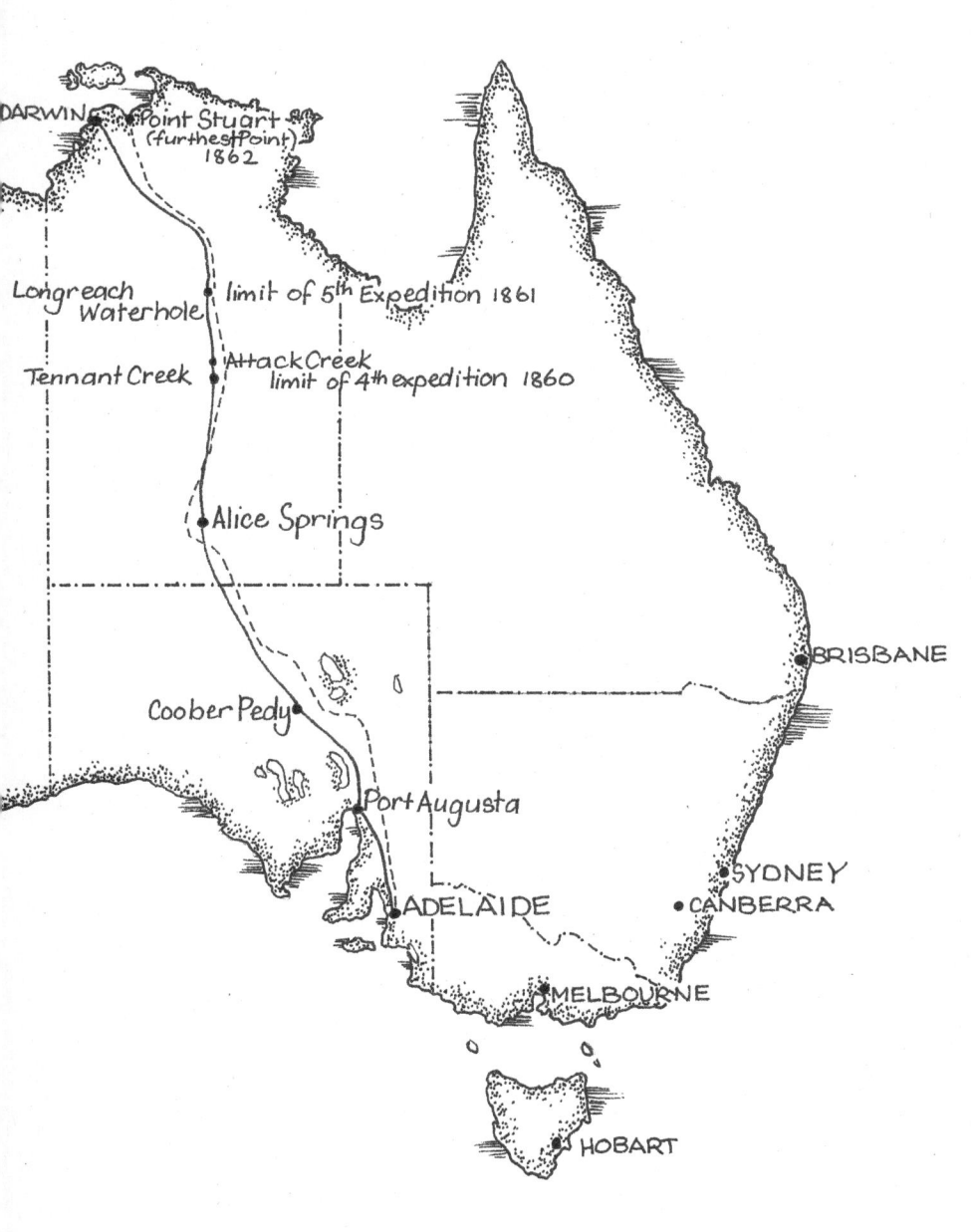

DARWIN Point Stuart
(furthest Point)
1862

Longreach
Waterhole   limit of 5th Expedition 1861

Tennant Creek   Attack Creek
limit of 4th expedition 1860

Alice Springs

BRISBANE

Coober Pedy

Port Augusta

ADELAIDE

SYDNEY
CANBERRA

MELBOURNE

HOBART

ONE

DEPARTURE DELAYED

The clink of happy hour invariably makes me smile. Tonight could be the exception.

Like a circling cat, I complete my second lap of the very campground I'd smugly dismissed half an hour ago. I settle on a patch at the outer edge, all grit and gravel and nudging the prison-grey corrugated boundary fence. My vision for this first night on the road had been a bush camp by a creek bed, not this suburbia on wheels. That was before the coffee cup incident, which I'm not ready to talk about.

Over at the shower block, a huddle of grey nomads grumble about knickers flapping on the washing line. One breaks off mid-moan to bark, 'There's no kitchen, if that's what you're after.' I'm not.

I scamper past another cluster who are nesting for the evening in front of a flickering television screen, stubbies and sensible plastic champagne flutes in hand, having far too much fun without me. Bet that cheery lot knows all about caravan park etiquette. If I could whistle, this is where I would trill a lilting jig and stride off into the sunset.

Setting up is painfully slow as I'm trying to be inconspicuous and nonchalant, faking experience for anyone watching. I place the lantern in the middle of the camping table and plonk down a plate of lettuce, tomatoes and olives. The portable stove's not getting a look-in. That lantern. It glowed so brightly in the camping shop. It's a right rookie giveaway, throwing a pathetic glimmer over my plate of salad that's already wilting—and not in a trendy hipster fashion. The waft of fried onions, my favourite outdoor cooking smell, drifts from the rig next door. Not helping.

I'm failing the grey-nomad initiation test, but that suits me fine. I'm not joining the club. Online forums are full of unkind comments about folk of my vintage who migrate up and down and around the continent with the seasons. Thousands at any one time. They drive slowly. Each evening at dusk, they gather to compare fuel prices, spruik camping gadgets and scratch around for something to grumble about. Like uncollected underwear.

My focus turns to the cask of a dubious red and I raise my tumbler to Bad-Ass Betsie, my strong and sensible four-wheel-drive. She won't take any shit. She has a bull bar in front in case I come in contact with a kangaroo and a covered canopy at the back, minimally retro-fitted with a plywood floor so I can sleep above ground. That's to avoid beasties of another ilk—the wee and not-so-wee creepy-crawlie variety. The yellow lamp on Betsie's roof is a relic from a past life down the mines, a sassy topknot hinting at a bold and brave existence. I reckon she's a feisty one and I fell in love with her at first sight: a no-nonsense mama, confidence without swagger. She also makes me look good, as the adjustable driver's seat means I'm not peering through the steering wheel. I could be mistaken for being tallish. Apart from Betsie, I'm travelling solo. I'm on a mission and have a deadline to meet. I want no distractions.

We both look out of place in this concrete carpark of a campground. Thank goodness I abandoned the hasty resolution about no alcohol on this trip.

I plan to follow the tracks of explorer John McDouall Stuart, the first European to ride through the centre of Australia from the south coast to the north and back again.

McDouall Stuart and I both hail from the Kingdom of Fife on the east coast of Scotland. He sailed on one of the first immigrant boats to South Australia in 1838. I was on one of the last (official) migrant boats in 1975.

He undertook six expeditions over four years, doggedly traipsing up and down the continent, getting a little bit further each time, looking more and more 'knocked up', a phrase he'd use for his horses. He was well aware that people lived on this land. He'd seen their tracks, their camps and other signs of habitation. But he wouldn't have had a clue that he was following trade and exchange routes created tens of thousands of years before any white explorer stuck a flag on these shores.

Me? One trip. One month. I'm already feeling a little knocked up.

Very little was known about Australia's centre in the 1800s. It's not that different in the twenty-first century. Vast expanses of the land I'm about to enter register zero population on the Census. Today, only three per cent of Australians live in the desert regions, which cover seventy per cent of the continent's interior. Most of us cling to the coast.

Within ten years of McDouall Stuart's final expedition in 1862, his route was used to construct the Overland Telegraph Line, a single strand of wire strung across 36,000 poles spaced eighty metres apart from one end of the country to the other. It's been dubbed the Internet of the nineteenth century, which tickles me, considering McDouall Stuart's reputation for being uncommunicative. The original Ghan railway line followed the same route, the first half from Adelaide to Alice Springs completed in 1921. All that's left of both are desolate structures still weathering outback extremes, broken slivers of ceramic insulators poking through the red soil and railway sleepers turning to sawdust. Later came the country's first transcontinental road, the Stuart Highway.

I'm using McDouall Stuart's personal maps and diaries to follow his tracks. I don't do maps well. Why did I consider this to be a sane idea?

My departure had been delayed for a couple of days as the rented satellite phone had been waylaid, flown right over my head from Canberra in the east to Perth in Western Australia. I suppose I should've been relieved it hadn't gone to Perth in Scotland. No worries, though. Australia Post promised it would land on my doorstep in Adelaide first thing this morning for an early start.

By 10 am I was packed and primed for action. I unfolded, perused and refolded my maps of the Flinders Ranges, Central Australia and the Top End. I didn't need a map of the first section out of Adelaide as it's a straightforward drive north. Preparation plus. And I waited.

By noon I'd culled the clothes bag and reassessed the need for multiple cooking utensils. Minimalism is my mantra. I was still on schedule.

By 1 pm I was in reverse gear. The fish sauce was back in the car fridge. You never know who you might entertain in the Outback. Did Melrose, 300 kilometres away, remain a realistic destination?

By 2 pm I was grateful for the delay as the hip flask filled with Drambuie had secured itself a spot. In the first-aid kit. I might get as far as Crystal Brook, 200 kilometres away.

By 3 pm I was so bored I'd shaved my legs. Tweezers for wayward chin hairs were in the toiletry bag, along with every conceivable weapon to repel insects, and, if all that failed, anti-itch lotion for bites. The Clare Valley, 100 kilometres away, could be the first-night destination.

I was sorely tempted to leave without the effing satellite phone but friends and family had made me swear not to leave without it. You don't need to travel very far north and the mobile phone is useless.

Then I heard the pfutt pfutt of the postie's motorbike. To misplace a brown-paper parcel is understandable, overlooking this monstrosity took some doing. The bright yellow safety box now stashed behind the front seat had heavy-duty snap locks and was emblazoned with multiple labels warning of a nuclear weapon inside, or so it looked to me. The dedicated SOS button on this marvel of modern communication technology was linked to a rescue mission centre in Houston, Texas. Fat lot of help that was in getting to my doorstep on time. I was off. I could read the instructions later. I had three hours of daylight.

All I had to do was drive up the Main North Road. This is no minor thoroughfare as the name suggests and yes, it heads in a northerly direction out of Adelaide. I couldn't get lost. The National Highway, the Australian National Highway and Highway One. Who would have thought they were different routes and different roads. I got lost, ending up on the one that's a 15,000-kilometre ring road around the circumference of the continent. It even jumps over the

Bass Strait for a quick jaunt around Tasmania. Goes nowhere near Australia's centre where I was headed.

Back on the right highway (already forgotten which one) and there was a baffling moment when I spotted the sun on the wrong side of Bad-Ass Betsie before realising I'd done a U-turn and was heading back to Adelaide. Then I got sucked into Gawler, this satellite town on the outskirts of Adelaide. Its claim to fame is that it was the first regional town in South Australia, founded in 1837. I'd no desire nor plans to go there today.

I'm facing the campground fence, pretending I'm alongside a creek bed in the Flinders Ranges. Toast to the moon. Ponder when I can reasonably slip into the back of Bad-Ass Betsie and stretch out in my swag. If I was at a table for one in a city restaurant, I'd pull a book out of my handbag to hide my Nigel no friends discomfort. I toast the full moon again and reach for my box of research papers.

Here's the old Australian Geographic magazine with its fold-out map and sketches of McDouall Stuart's expeditions. That's what started this whole escapade. He kept pushing himself into my life. I'd never heard of him until that magazine fell at my feet in a second-hand bookshop in Adelaide in the 1980s. News of McDouall Stuart's successful crossing, which hit the press worldwide back in the mid-1800s, was as important as the first moon landing in 1969.

I found myself slipping his name into conversations as you do when a new lover comes along. John McDouall Stuart. Difficult name to say without affecting just a tinge of a Scottish brogue. Easy name, however, to confuse with Captain Charles Sturt, the Father of Australian exploration, the man who gave McDouall Stuart the exploring bug in the first place.

Folk around the dinner table to this day frequently get them mixed up. He doesn't get a mention in some Australian history books. What's more, those who have heard of McDouall Stuart lower their eyes, shake their heads and mutter 'Bit of a pisspot you know'. He could do with a good PR agent.

What possessed this wee man from a dreich Scottish fishing village to venture again and again into the hot, dry, dusty arid lands of the Aussie Outback? Was he looking for fame? Hope not. Look at Burke and Wills, who were exploring at the same time. Tragedy has made that pair Australia's most memorable exploring duo. The spectre of Burke and Wills kept surfacing, side-lining his achievements. It still does to this day.

Respect? The media gave McDouall Stuart the illustrious title of Prince of Explorers, but acknowledgement from some politicians in the fledgling South Australian Parliament was lukewarm. He didn't seem that fussed, mind you.

Fortune? That didn't happen. The South Australian government was stingy with its rewards and he couldn't squeeze a measly pension out of the government over in Britain.

A chance to dry out? This is a theory that's been put forward to explain why he'd return to Adelaide, down a few bevvies, and head back out again. Surely there are easier ways.

He wasn't a great catch, it's true. A fine figure of a man he never was. He didn't age well either. The effects of scurvy from seriously deficient eating habits can do that: spongy gums, bad breath, loosened teeth, protruding eyes, blackened skin. When questioned about the absence of a flattering photograph, his family is forced to respond with a 'no comment'.

Nice eyes though.

Okay. The coffee cup incident. You'll find out anyway so I might as well tell you now.

It was the skittering that got my attention. There was I, in the Gawler supermarket carpark, scouring the map spread across the steering wheel, scanning for camping areas on the outskirts of town, when that takeaway cup hopscotched across the parking bays and skidded to a halt against the front tyre of a white van skulking under a tree. A scruffy shadow lurked behind the steering wheel.

At my farewell knees-up in Adelaide, I stuck my open hand out in defiance whenever 'Peter Falconio' or 'Wolf Creek' were mentioned. I had not watched the documentary or the horror movies based on Falconio's murder along the Stuart Highway and, no, I was not interested in any updates. 'They never found his body.' 'It happened on the same road you're about to drive on.' Did I listen? No.

I shoved my road map out of sight and snuffed the interior light. A newspaper headline flashed before me: 'Granny's body found 40 km from home.'

The police would search my home for clues. I should have tidied away those months of planning before I left. The pile of clothes slithering out my bedroom doorway would give them a false impression of a hasty exit. What was she up to? Or a forced removal. Who was after her? But no, the map of Australia on my bedroom wall would tip them off that I had a plan, and the plan was to get much further than this town. My route is marked, a thick red line snaking up through the middle of the continent, pencil lines leading to post-it notes listing sights to see, photos to take, people to meet. Meetings I would now fail to turn up for is what I was thinking. The doomsayers were right. I was doomed. People disappear... bleached bones... shallow graves... dental records. These were the phrases worming their way into my brain. Once heard, hard to unhear. I hadn't bid a decent farewell to my kids and grandkids was all I could think of. Then the driver returned to her van to the delight of her scruffy dog, sitting behind the steering wheel. I'll phone home tomorrow.

Please don't ask me why I'm doing this trip. To prove something? To learn something? Maybe both. Maybe more. Maybe that's all a load of bullshit. Following McDouall Stuart's tracks has given me a purpose to head off on this adventure, but is he the real reason or an excuse?

The explorer title didn't particularly impress me. Swags of bearded men were crossing left and right across Australia in the nineteenth century. Their interweaving lines on explorer maps are so fankled, it's a surprise they didn't keep bumping into each other. Instead, they had a careless habit of losing their way, losing each other, and losing their lives. And while we're at it. What the hell am I doing, writing about a dead, white male explorer? The last person I thought I'd choose to follow across the country. Shouldn't I be writing about a woman explorer at least?

A snippet among my notes says that, on the final expedition, McDouall Stuart's men had come through Gawler. I'm cheering up. They would've camped near where I am now. Another toast is called for. I'm feeling like quite the explorer myself.

I can hear McDouall Stuart now. Just as I was starting to feel good about this trip. *Feart fur a dug, hen?* he'd say. I've no idea, but I fancy him talking this way, a dollop of Fife, as that's where he grew up, a wee bit of posh, pan-loaf Edinburgh, as he lived there for a spell as a young man, and some West Coast banter from the area his parents came from. He was said to be a fan of Robert Burns, the bard who brought the Scots language to the world's attention, so that's that.

If he was here, I'd tell him straight. 'You've been pestering me for years,' I'd say. 'Goading me to retrace your crazy expeditions,' I'd say. 'And now you've got the cheek to be having second thoughts?'

A throat-clear from my campground neighbour. I may have said all that out loud.

TWO

A NEW DAY DAWNS

The crack of dawn. The crunch of gravel. So much for slinking out before the happy campers surface. I'm scrunched into the foetal position in Betsie's rear, discovering that a space one centimetre too short might as well be ten. Caravans file past as I unkink my spine. I do hope I'm steaming up the curtainless windows.

Flinders Ranges, here I come. Travel writers lose all sense of decorum when it comes to describing the mountainous spine swirling into the horseshoe-shaped amphitheatre of Wilpena Pound. Especially after they've taken a scenic flight to admire its grandeur and recognise landmarks depicting the Creation stories passed down by the ancestors of today's Adnyamathanha Nation. The superlative goes into overdrive: every colour is heightened and every metaphor is squeezed out to report the ancient landscape as 'our unspoilt natural environment'. They can't help themselves.

Early explorers, however, were oblivious to its charms. To be fair, they were distracted. In 1802 Captain Matthew Flinders had a bit of a look around when he sailed the *HMS Investigator* up Spencer Gulf. He wasn't particularly taken with what he saw. Probably wouldn't care a jot that the ranges were later named in his honour. 'Dead uninteresting flat country everywhere,' he said in his journals. He named two high spots—Arden for the one he thought was the furthest peak, in honour of his granny; Brown, the highest peak in sight, named after the ship's botanist, Robert Brown. Now there's a man who took plant spotting seriously.

Brown also had little to say about this region, too busy plucking petals and leaves off thousands of plants. He's the Brown who gave the name to Brownian Motion, which high-school science students learn is all about the movement of atoms and molecules. In my view, however, Brown's career high point was his inspired choice of name for one of the many grasses he spotted on his Aussie trip: *Panicum effusum*. Hairy panic if you want to be common. To this day, there is indeed *Panicum effusum* as these gigantic fluff balls roll through the streets of country towns, barring doorways, covering windows and sometimes reaching roof height. I'm hoping it makes an appearance on my trip if only to have the chance to announce: 'Is that hairy panic I see before me?'

After Flinders and Brown came young Eyre, who was equally unimpressed with the terrain. The whippersnapper was a mere twenty-five years old when he set off in 1840 on his Great Northern Expedition to get to the centre of Australia. He saw these ranges as obstacles in his way. Eyre climbed a mountain and, so disappointed was he at what he saw beyond, he called it Mount Deception. He named Mount Hopeless even before he climbed it. The view from the top, he said in his journals, wasn't just hopeless, it was positively cheerless. Mount Distance showed he was finding this all somewhat tiresome and Termination Hill speaks for itself.

Meanwhile, those Adnyamathanha ancestors went on with their lives as they'd done for eons, like they owned the place, which, of course, they did and do. They didn't need today's drones and aerial photography to appreciate their surroundings.

Then came the pioneer pastoralists who lost no time in creeping up north of Adelaide and through the Flinders Ranges. And so South Australia's own squattocracy was born. Before his exploring days, McDouall Stuart was one of the busiest surveyors in the region though he didn't seem that interested in getting a piece of land for himself.

I untangle Gawler's reef-knot of exit routes and head north.

The sign over the deserted railway platform says I've reached Mount Bryan. No amount of twirling of the map on Betsie's bonnet alters the fact that I've drifted east, heading for the Victorian border. This tiny township is named after the would-be explorer, young Henry Bryan, a cheerful lad and an optimist, who landed in Adelaide in 1838, around the same time as McDouall Stuart. Within a year, during an expedition around here, he got lost, never to be seen again, his riderless horse wandering back into Adelaide several months later. I feel for him.

McDouall Stuart had little need for maps. He was the one creating them for others to follow later. His party trick was to halt his horse, take a puff on his pipe and announce that a landmark he'd pinpointed on a previous expedition was within cooee. And there it would be. Every time. He used only a prismatic compass and a pocket watch. Smart arse.

I can tell what he'd be saying if he was around to witness me turning my map this way and that. *'Think this is gaun tae be a doddle, hen? Are ye gaun tae get lost, is what I'm wonderin.'*

Within cooee for me is Burra, a two-hour jaunt in the family car from Adelaide. If you don't get lost. Burra was once the copper capital of South Australia, with one very productive mine aptly named the double-barrelled Burra Burra. With a population heading for 5,000 in its heyday, it was the second-largest settlement in South Australia after Adelaide. Burra is now a historic tourist town, with a historic square, historic town hall, historic open-cut mine, historic miners' cottages and multiple museums. Might as well visit while I'm in the area.

'You've got an accent.' The name tag tells me I've met Geoff, a volunteer historian in a museum on the main street. Geoff is keen to chat about McDouall Stuart, so I smile sweetly.

'Chambers and Finke took a chance backing him, y'know. He was reckless.' Here we go.

I resist the temptation to tell Geoff about the dodgy deals I've read involving the two men who were to bankroll his expeditions. William Finke, a skilled geologist, was even more skilled at creative land deals. He was on the lookout for lucrative mining leases. James Chambers was hell-bent on creating a pastoral empire. McDouall Stuart loved going bush and was happy to do their bidding.

'Many adjectives describe McDouall Stuart but never reckless.'

'C'mon, you have to agree...he was a bit of a drinker.' Geoff winks. 'You Scots. You can't help yourselves.' I glare into the eye that winked.

'Do you know why Adelaide's where it is?' I've no chance to say yes or no. 'It's far enough away from there,' Geoff cocks his head towards the Victorian border, 'So's nobody would try to come over here.'

Interstate rivalry between South Australia and Victoria began when the border lines were first drawn and continues to this very day. There's a squiggle at the top of the straight line near the junction of South Australia, Victoria and New South Wales—a measuring glitch from the 1800s—and South Australia has been grumbling about being diddled ever since.

Forget about Australia becoming one united country in 1901. On board the boat coming out, I was assured by each Australian I met that I was making a terrible mistake in not heading to wherever they came from.

My favourite brawl between South Australians and Victorians is from the 1990s, amid accusations that Victoria had 'stolen' the Grand Prix from them. Egged on by their political leaders, they were on the verge of boycotting each other's beer. It got that serious.

'We're a purer pedigree than that lot.' Geoff's on a roll. 'South Australia is the only free State, y'know. Free from the convict stain.'

I've heard that before too.

South Australia is where the term 'province' was preferred in certain circles to set it apart from the colonies populated by convicts in the eastern seaboard and Tasmania. This was to be a beacon of respectability.

The British government had become twitchy about populating Australia with its overflow of criminals, many bolting into the bush to escape. Free settlers weren't following the rules either. Cashed-up colonials with a sense of entitlement became illegal squatters, snaffling up land outside the surveyed borders and creating massive country estates for themselves. Rumours circulated of massacres and poisonings after clashes with First Nations warriors defending their land and their women, incidents rarely reported in the media. Those illegal squatters, wielding increasing influence in high places, became a class of their own—the 'squattocracy'—and their unseemly land grab was unstoppable.

Systematic colonisation was the counterplan for South Australia. The Wakefield plan, it was called. Unlike the eastern states, there would be no convicts here and land would be properly surveyed and sold. Had no one noticed that Edward Gibbon Wakefield penned his plan as he sat in a prison cell in London? Lucky for him that his crime was kidnapping a wealthy heiress rather than nicking a loaf of bread. Wakefield gave his plan the title, 'A Letter from Sydney', pretending he was living in Australia—possibly to give his idea a more authoritative ring about it than 'A Note from London's New-gate Gaol'. A key feature had been to price the land out of reach for those on assisted passage to ensure cheap labour for the aspiring landed gentry. This was to counter the lack of convict labour. Natural slavery was Wakefield's phrase for his idea.

What could go wrong?

Power dynamics, double standards and mixed messages plagued the start of the new colony. As well as the first group of settlers, the ships bound for South Australia in 1836 brought over two powerful groups of men with conflicting intentions: one consisted of government officials to run the colony and the other comprised Colonisation Commissioners, here to sell the land to make money to fund this venture.

The story goes that the settlers were gathered together for a ceremony under a gum tree in what's now the seaside suburb of Glenelg, with the first governor, John Hindmarsh, reading out the Proclamation of South Australia and Colonel William Light, the province's first surveyor general, by his side. This version is backed up by a very large painting owned by the Art Gallery of South

Australia, considered a work of historical significance.

The truth is, Colonel Light did not attend the ceremony, stating he had better things to do, and Governor Hindmarsh neither wrote nor was the one to read out the so-called Proclamation. To this day, that speech has been given weight it doesn't deserve. It's been called a Founding Document for South Australia but has no legal standing. Some say the speech was cobbled together on board one of the ships sailing into shore; others that it was scribbled on a piece of paper inside one of the tents. By the time the document was printed as a poster, it abounded with capital letters to give it an air of supreme importance.

The men of power knew what they were doing. They weren't about to raise the issue of prior land rights in this speech. So no mention of the Letters Patent signed by King William IV. It's a legitimate founding document and outlines the rights of the existing inhabitants over their lands. No mention of the Colonising Commission's own report that land would have to be ceded by the existing occupants before being sold. That report had been published in London before they set sail. No mention that the former governor of Van Diemen's Land (Tasmania), George Arthur, implored South Australia to sign a treaty with the original inhabitants to avoid the mistakes he'd made.

It would be a hundred years before those Letters Patent got any attention in the corridors of political power.

The settlers instead were told to behave and conduct themselves with 'order and quietness' among each other and towards the original inhabitants.

That painting, produced twenty years after the landing of the ships, is now known to be a work of imagination and it's clearly of a genteel scene before the kegs of rum were opened.

Journals written by folk who were there on the day speak of sailors getting so drunk they couldn't board their ships, carousing settlers partying not only into the night but well after the break of day, and at least one of those men of power, Colonial Treasurer Osmond Gilles, getting horribly drunk. Building a gaol was considered unnecessary. There's optimism.

By the time McDouall Stuart arrived two years later, those men of power were the ones ignoring the plea for order, squabbling among themselves over who was really in charge. Governor Hindmarsh had gone and Robert Gawler, who took over, was himself relieved of his duties after a couple of years. Colonel William Light, South Australia's first surveyor-general, who had decided he was 'too busy' to attend that celebration, had resigned in a huff and the next two surveyors-general each lasted less than a year. Colonial Secretary Robert Gouger was suspended from his duties after a public brawl with Treasurer Gilles. All bordering on a theatrical farce. The systematic pegging out of the land was in a bit of a shambles.

Keeping South Australia pure of the convict stain wasn't a roaring success either: convicts persisted in sneaking over the border. Not only a gaol was built, but gallows as well.

The Kaurna Nation, population around 700 in 1836, was the first on mainland South Australia to feel the blow of dispossession. The heart of their country is where the city of Adelaide now stands and Karrawirra Pari (red gum forest river), now called the River Torrens, was a major resource and a focal point for ceremonies. An early map of Adelaide shows the Kaurna relegated to a small section on the northern banks of the river, now called Pinkey Flat.

Neither Wakefield nor King William was around to see their

wishes sidelined. Wakefield never bothered to set foot in Australia, far less South Australia. You'd never think it, considering the streets, buildings and landmarks bearing his name. And King William died shortly after he named South Australia's capital city after his wife Adelheid Amalie Luise Theresa Carolin of Saxe-Meiningen—thankfully abbreviated to Adelaide. The Kaurna people call it Tarndanya.

McDouall Stuart's arrival in Port Adelaide was at a peak period of assisted passage. Thirty-four ships set sail from Britain in 1838. Steerage-class passengers cramped below the water line contended with lice in their hair, weevils in the flour and rats eating their boots. Deaths at sea were common.

His accommodation on the good barque *Indus* was above water but McDouall Stuart wasn't a good sailor and his weak constitution had him regularly heaving over the bow of the ship. After four months at sea, he would have been relieved when they docked in Port Adelaide in January 1839.

My biggest worry on board the *Australis* wasn't lice. It was ice for the gin and tonics to numb the boredom of playing Monopoly every day for five weeks. I discovered I had a cast-iron stomach, accepting extra helpings of lima beans as, one by one, seasickness meant dinner table comrades failed to turn up.

This was 1975 and assisted passage was in its death throes. More than a million migrants, mostly from the UK, had arrived since the Second World War, thanks to the Big Brother and Populate or Perish schemes and the Bring out a Briton campaign. We were all lumped under the label of Ten Pound Poms, even though we Scots to this day assert that only the English are Poms.

McDouall Stuart's Adelaide had around 6,000 inhabitants. Homes at the start were tents and huts with walls made of mud and rushes, earthen floors and tarpaulins for roofs. Permanent homes, still with earthen floors, had walls of limestone, timber and rammed earth and roofs of slate, thatch or shingles.

When we arrived, Adelaide's population was around 800,000, scorned as a big country town by Sydneysiders and Melburnians. It still is. I saw Adelaide as a cultured, socially progressive city. And clean.

McDouall Stuart had no problem getting a job. His skills as a surveyor were immediately in demand as the land was not being pegged out as quickly as promised for settlers keen to settle.

We, too, were keen to settle when we arrived, embracing the Aussie dream of owning a triple-fronted home on a quarter-acre block.

Any job would do us. It was part of the adventure. 'Downward occupational mobility' was the official phrase befalling the average migrant, but I was young and on an adventure. I became a temp secretary, with above-average shorthand from my days as a cub reporter for the local newspaper back home but with questionable typing skills. I lasted long enough in each job for work colleagues to teach me words like smoko (tea break), crook (off work sick) and—Durex. Where I came from, this was a brand of condoms (here it was adhesive tape) so it was disconcerting when I was asked if I had any to spare because the entire office supply had been used up.

I loved discovering new foods and watching Australian Rules football (in blissful ignorance of the rules). Footy on a Saturday was where I learned the delights of the Chiko Roll. Then again, I also like haggis. I embraced different: the spindly branch masquerading as a Christmas tree wilting and turning grey overnight, the gravel scrapes instead of putting greens on rural golf courses, even the swarms of flies settling on our backs, turning white T-shirts black.

My letters to Scotland in the early days contained photos of me squinting in the bright sunshine, pulling oranges off a backyard tree. Imagine having an orange tree in the back garden, walking out there, picking one to peel and eat. This was the life.

I never wrote home when I was homesick.

Geoff leads the way out of the museum to show me a plaque on a nearby wall, proof that McDouall Stuart wuz here. On his sixth and final expedition, he stopped in Burra to send a telegram to Adelaide to say he and his men had returned safely. 'Fancy that,' I say.

I'm sure explorers of note never kept to a straight path, forever straying and making their own wee discoveries, muttering 'fancy that'. All part of the job description. I'm starting to think I could be channelling the man.

PART 2

FLINDERS

Surveying Years and
Exploration One
May – September 1858

RANGES

from Adelaide
to Marree

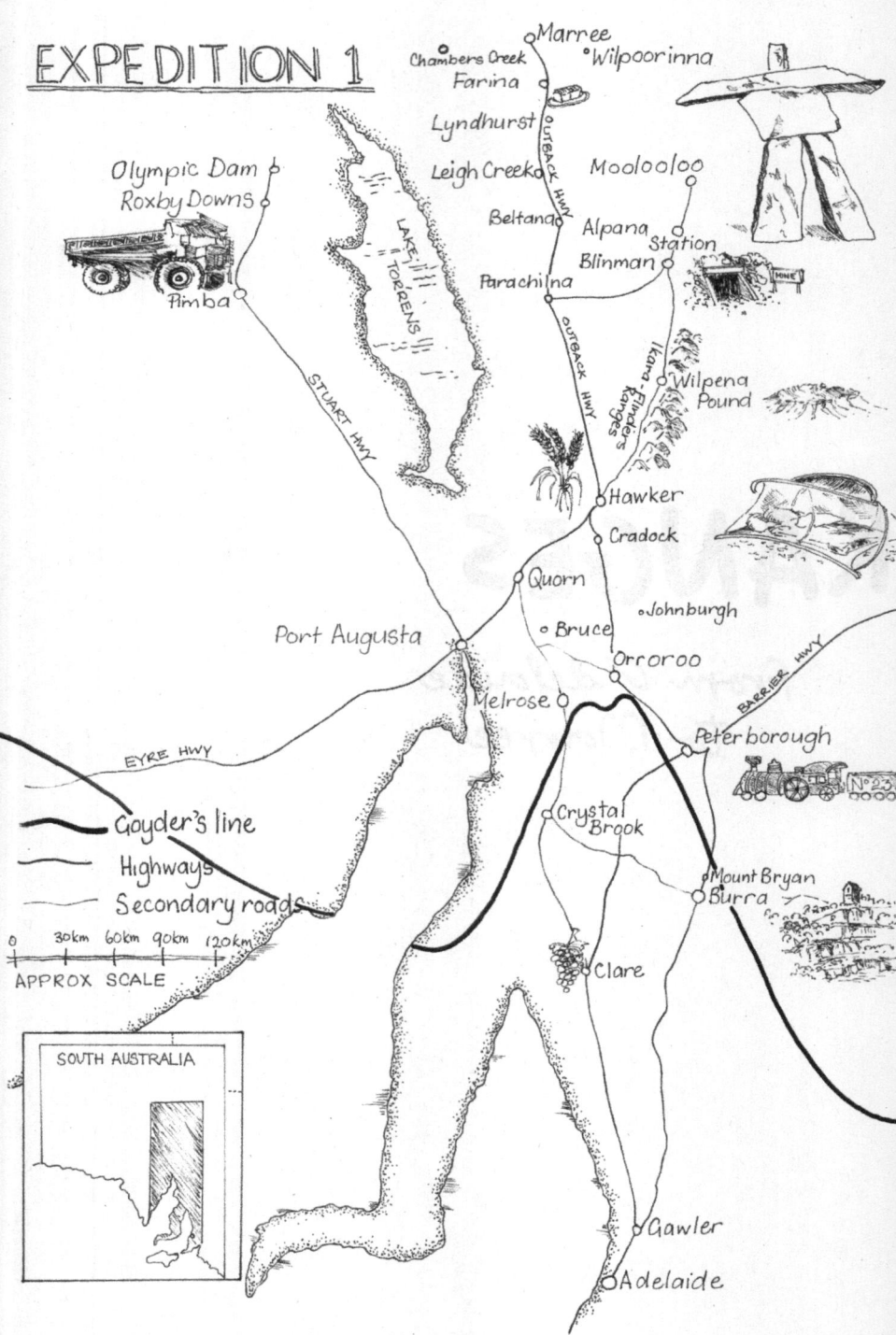

EXPEDITION 1

Marree
Chambers Creek
Wilpoorinna
Farina
Lyndhurst
Olympic Dam
Roxby Downs
Leigh Creek
Moolooloo
Beltana
Alpana
Station
Blinman
Pimba
Parachilna
Wilpena Pound
LAKE TORRENS
STUART HWY
OUTBACK HWY
Ikara-Flinders Ranges
Hawker
Cradock
Quorn
Johnburgh
Port Augusta
Bruce
Orroroo
BARRIER HWY
Melrose
Peterborough
EYRE HWY
Crystal Brook
Goyder's line
Highways
Secondary roads
Mount Bryan
Burra
0 30km 60km 90km 120km
APPROX SCALE
Clare
SOUTH AUSTRALIA
Gawler
Adelaide

N°23

THREE

THE CREEP OF PASTORALISM

Something's odd as I scoot north from Burra, barrelling up the Barrier Highway towards the Flinders Ranges. I'm one of those side-alley clowns you pop balls into at the fairground; mouth open, head swivelling.

Drift left and I'd glide through swaying fields of blond wheat, past a shiny tractor glinting in the noonday sun, ready for action—a scene from one of those propaganda films enticing migrants like me to Australia. *There is a land where honey flows / Where laughing corn luxuriant grows* are the lines from Song of Australia that spring to mind.

Veer right, however, and I'd stumble around stony outcrops, straggly trees and scrubby saltbush. A crumbling chimney in a paddock of dust and dirt points accusingly at the cloudless sky, more reminiscent of another poem written in the day by John O'Brien, with the oft-repeated refrain, *We'll all be rooned' said Hanrahan / Before the year is out.*

I'm straddling the Goyder Line, used to determine which pastoral lands were eligible for the first, but not the last, drought relief package for the region. It's not a straight line, but a wavy sweep drawn across the map of South Australia back in 1866 and it has materialised before my eyes.

Debate continues today about how the state's surveyor general at the time, George Goyder, decided on that looping line. Was it rainfall and, if so, what level of rainfall? Was it the vegetation? I have a bitumen road to follow.

Passing through this region over the years, on the way to somewhere else, I was seeing a landscape that's been the same since time began. Or so I thought. Truth be told, in the space of 200 years, just a blink in history, so many mistakes were made it's said much of the landscape is unlikely to ever recover. So much for being unspoilt.

In the 1840s, less than a decade after the first boats arrived in South Australia, there were sheep stations past what is now Hawker township, 400 kilometres north of Adelaide. To participate in this game, it was cash only, no credit, so a pastime for the rich.

All the aspirational pastoralists had to do was get a surveyor like McDouall Stuart to peg out a parcel of land, put some stock on it and hot-foot it back to the Lands Department in Adelaide to pay for a lease. For a few years, they were laughing, thanks to a system called special surveys, which allocated additional land to people who bought plots in Adelaide. Huge tracts of land were taken up by absentee landlords, many with no intention of settling there, interested only in land speculation. They would cherry-pick the extra land they wanted, near rivers and permanent waters, often in such a way that land surrounding their parcels couldn't be used and so that fell to them too. Very handy. Sixty per cent of the special surveys handed out went to folk from outside the colony. The system was abandoned when it was realised what was going on.

In the following ten years, nearly all of the land thought to be suitable for grazing was occupied up to and past what is now Leigh Creek, more than 500 kilometres from Adelaide. Never a thought to acknowledge the land rights of the original inhabitants. And the idea of asking them for advice was beyond the ken of most of the newcomers. The colonists gambled with an environment they didn't understand, perplexed by the strange plants and animals they came across, the lie of the land, the poor condition of the soil, the flow of the water. Nothing but saltbush was a common complaint with no notion of how nutritious it was. Today, saltbush lamb is on the menu in the best of Adelaide restaurants.

Then came the Great Droughts.

Records show that one-third of the sheep and half the cattle perished. Many hopeful pastoralists went broke, leaving the wealthiest, those with the resources to last the distance, with the chance to pick up leases at bargain prices. First Nations people, who'd lived here for thousands and thousands of years before being pushed off to the boundaries of their land, had already been decimated by diseases like smallpox, measles and tuberculosis. They were now dying of starvation and thirst, denied access to waterholes, and their native food sources wiped out by the years of pastoral activities.

'Mention the Goyder Line today, and you can still start an argument.' I'm chatting to the owner of a coffee shop in Orroroo, 280 kilometres from Adelaide, where an installation on the edge of town tells me I'm at the northernmost tip of the northernmost arc of the Goyder Line before it swept westwards from here.

'Two-faced country.' That was to become George Goyder's view in later years. Harsh. But who could blame him? What thanks did he get after riding thousands of miles on horseback, choking on dust storms and battling hot winds? Goyder's Line of Foolery: that's what folk called it.

Goyder's big problem was that he went beyond the task allotted to him to decide who should get drought relief. He went on to advise that the land north of the Line should only be used for light grazing. Never for crop farming.

Courageous move, Mr Goyder.

The outlook as Bad-Ass Betsie and I drive beyond the Line is rolling hills and stony outcrops in all directions. From here it's semi-arid country—hot summers and cool winters and rainfall that's reliably unreliable. Land that can only cope with one sheep per 3–4 hectares. Who was the clever dick to promote the idea of cropping?

'Rain will follow the plough' was the counterargument to Goyder's edict that farming should never be allowed above the Goyder Line. This baffling concept was imported from the United States where it started as superstition and gained pseudo-scientific status. We'd never make that mistake today, would we? The basic premise was that churning up the soil could goad Mother Nature to increase rainfall.

The drought had ended and the pastoral land was looking good again. The government had a problem; they were running out of land and the influx of new arrivals were tempted to look interstate. The government looked north and had a bright idea. The Goyder Line was about to be ignored. Between 1869 and 1884, the government took back two million acres of land from the pastoralists—8,000 square kilometres—and turned it into small farms.

The intention wasn't just for an agricultural boom. Oh no. South

Australia was to be the wheat belt of the nation. The government gazetted several townships to service the farms and laid down the tracks of the Ghan railway to transport all that wheat and barley. Some of the names on the Flinders Ranges map lying on the passenger seat beside me reflect the background of those thousands of families who grasped the chance for a slice of Australia and headed north; Melrose sounds Scottish, Clare sounds Irish while Crystal Brook sounds very English-countryside. The German influence is hard to pick, with certain town names eliminated when anti-German sentiment spread throughout the country after the First World War. Petersburg, for example, was changed to Peterborough.

Farmers and their families, second and third-generation settlers, arrived with high hopes of success, so wanting to believe that rain would indeed follow the plough, only to face a tough life. Eventually, they gave up, walked away from their dream farms and the land reverted to pastoral use.

Cradock, destined to be one of the many towns to service the fantasy, if only rain had followed that damned plough, has a population of thirty-five and a great little pub once called Heartbreak Hotel. Carrieton's top attraction is the road to Johnburgh, says a tourist brochure. Johnburgh itself is described as a 'fascinating example of a town that died'. Then there's Bruce. In Australia, a Bruce is never far away. This Bruce is a town without people and a railway station that's never open. Many gazetted townships never became more than a name on the map.

I walk along dry creek beds and struggle to clamber over fallen trunks of huge trees, uprooted by raging torrents. The surrounding land is flat except for deep cuts gouged into the earth. A legacy of flooding waters or devastating drought? No one's around to ask. Skinny trees, blackened, with spikes lined up along their spindly trunks, stand in clumps. Bushfires, I'm thinking.

A chimney, almost intact, rises from a sturdy homestead and I'm sold; transported to the back door scattering seed for the chooks, inside the country kitchen baking bread, out the back tending the veggie patch in wellies and headscarf. Good job a real estate agent isn't handy to take advantage of my pipedream. I've never fed a chicken, can't bake bread and I kill mint which is said to be unkillable.

The blank windows on either side of the front door stare out like an old biddy waiting for something to happen. No one's been home since the family packed up a century ago and walked out that front door for the last time. Rows of fencing wire, once taut, slump in submission between redgum posts that once stood erect. No wonder the occasional sheep roving over the landscape is baa-ing in mirth, the crows joining in with their mocking caack-caack-caack.

A picturesque ruin, a lonely tree, wrecked windmills against the skyline, stone tanks the worse for wear—all vying for attention from an artist or photographer to come along and make them famous, or at least frame them to hang in their hallways. The untold story outside each frame is not a happy one.

I pop out at Hawker township, which once had two flour mills as it was expected to become a major service centre for that anticipated wheat belt. The town has transitioned into the tourism gateway to the Flinders Ranges. Surrounding pastoral stations, too, rely on tourism as their saviour, helping station owners make a decent enough living off the land.

I see now that the chimney stack was giving me the finger.

Still, I hold onto the notion that I'm looking at an untouched landscape. Why do I feel this way, even when I know better? Sucked in by the tourism marketing sentiments asking us to enjoy nature 'in the pristine state Mother Nature intended'; distracted by the geological data about mountain peaks created 300 million years ago; mesmerised by ancient Aboriginal Creation stories of the formation of creeks, peaks and waterholes.

'The bones of nature' is how Australian landscape artist Sir Hans Heysen described the Flinders Ranges, a region he returned to again and again to create his drawings, watercolours and oil paintings. I get what he means when he wrote that 'everything looks so old that it belongs to a different world'.

I've seen paintings by colonists in the early days, of landscapes that look like parklands. I thought they'd maybe subconsciously transposed their English countryside onto the Aussie outback. I've read journals like McDouall Stuart's, also speaking of wooded areas that looked planned and of land that reminded colonists of the country estates they came from. That's not what I'm seeing now.

Up the road is Ikara, where rock faces, millions of years old, glow pink and scarlet and red in the setting sun. The early pastoralists who saw this natural amphitheatre called it Wilpena Pound, pound being an old English term for animal enclosure.

The Adnyamathanha creation story for Ikara revolves around two serpents. A basic version is that the serpents travelled south, formed the ranges, then formed a circle and willed themselves to die, creating Ikara, which means 'meeting place'. Ikara's highest peak, Ngarri Mudlanha / St Mary's Peak, is the head of the female serpent.

Ikara is within one of the earliest pastoral stations in the region, dating back to 1850 and a concentrated example of how the Flinders Ranges has been used since colonisation. It became a sheep and cattle station. Later it became an experimental wheat farm, at one stage it was declared a forest reserve, and then its tourism potential was recognised.

I feel I'm in the confessional. I swallowed the notion that the First Nations people in South Australia were a compliant lot: walking away when the newcomers arrived; letting them install their sheep and cattle and take over the permanent waterways; seeing the native game disappear; standing by as the water resources were spoiled and the land mismanaged. Much meeker than the Maoris in New Zealand, we were told.

I also accepted First Nations people had little connection with the land: their portrayal as nomads and hunter-gatherers embedding that view; the term 'walkabout' bandied around, a cute but loaded word. Some explorers' journals, like McDouall Stuart's, did mention they came across examples of tools and contraptions for tracking and trapping, but the idea of land management would have been laughed at. I was certainly ignorant of Aboriginal people's knowledge about sustainability and plant diversity, developed over hundreds of generations of trial and error. So much more than setting traps, working out which plants were edible and using bird flight and the lay of the land to source water.

I've since read about the disconnect when it comes to land-owners talking about settler–Aboriginal stories. Encounters, friendly or otherwise, were always with other people, in other parts of the country, becoming hearsay, told almost like fairy tales, Grimm or otherwise. A disposition to turn away from pre-colonial history, except perhaps the 'safe' zone of Dreaming stories, brush aside the culture, the languages, the knowledge of the First Nations peoples who have lived here for tens of thousands of years. To shrug off the idea of resistance, of frontier violence. I wonder where that comes from. Is it similar to latter-day migrants like me, who find it difficult to accept the idea that the society I came from the other side of the world to join, has a flawed history? We all want to feel at home, to have a sense of belonging. And we want a rosy story, to feel comfortable.

It would be so convenient to keep saying that we weren't told what was going on. But there are so many written accounts, in newspapers, government records, letters and private diaries, it's impossible to deny. Clearing up that myth of peaceful take-over is a University of Adelaide project 'The South Australian Frontier and its Legacies' which uses a digital story map to illustrate how, as settlers spread out from Adelaide in all directions, conflict broke out between the colonists and the colonised.

The next stop for me is Blinman township. It sneaks up on you, a main street with a smattering of buildings.

Not only pastoralism but also mining changed the face of the Flinders and the story goes that Robert 'Peg Leg' Blinman started the copper rush here in 1859, when, bored witless overseeing a flock of sheep, he absent-mindedly chipped at a rock he was lounging on, exposing a streak of green. Suddenly not so bored, Peg Leg hopped off that rock, rounded up some mates and pulled together £10, a year's salary for a shepherd, to buy the mineral lease, paying for a year's rent.

Word got around that there was copper in them thar hills and more than 150 mining leases soon dotted the landscape.

A hotel came first for Blinman's hot and thirsty miners. Families arrived, so a school was built. A post office, then a general store followed. In the early days, children played in the swirling dust blowing over a tent city that sprouted beside the opening to the major mine shaft. Their dads coughed and choked underground while their mums and older siblings walked endlessly along dusty bullock tracks gathering wood and water to feed the mining machinery.

While Burra had the biggest mine in South Australia, Peg Leg's became the largest and most productive copper mine in the Flinders Ranges. After he sold it.

In its heyday at the turn of the twentieth century, Blinman had a population of 1,500, large enough to boast a Blinman North and a Blinman South. Today, the population is nudging twenty and the one remaining mine is a tourist attraction.

I head for the general store, which offers quandong ice cream and phone cards for the public booth up the road. My mobile phone is reduced to a glorified alarm clock and I want to ring home. A click and 'please leave a message' but no one can ring me back. Will they be wondering where I am, what I'm doing?

I'm here to visit the cemetery at the far end of the main street, on the edge of town. Somewhere among the headstones is the grave of the man who was to become McDouall Stuart's second-in-command and possibly the most loyal companion anyone could wish for.

The chain on the gate clanks against the rusty railings as I drag a shallow semi-circle through the red gravel, the scraping of metal nearly drowning out the whirr of flies circling my head. My furious flapping makes no difference as I walk around headstones that are leaning into each other like punters sharing a yarn at the local pub. A few have hit the ground, face down. Inscriptions on still upright stones tell fading stories of a tough life in a community that was far from the attention of any authority, medical or otherwise. Lung disease was a common cause of death; so too, mothers dying in childbirth.

I crunch past clumps of weedy floral tributes among the graves. The cloud of flies still hovers. The inscription on a very old stone is in memory of a beloved wife who departed this life in 1874, aged twenty-nine, and her twin sons. Some graves are not quite so old. On top of a pile of stones is a broken piece of slate, a simple memorial scored into the surface by a pal to commemorate Len Mac Macintyre, born 1916 – died 1957. Very pretty weeds have sprouted around it.

And then, there it is. Sparkling white in the noonday sunshine, a splendid obelisk marks the grave of William Darton Kekwick. A plaque states: 'In admiration of his pluck, endurance and loyalty to John McDouall Stuart, as second in command from 1859 to 1863.' The obelisk was erected by the Royal Geographical Society of Australasia SA branch and friends in 1909.

'*Wondered when ye wur gaun tae spot it.*'

'Who said that? Shit...shit.'

'*Language hen. A didnae let ony o' ma past companions swear. It wis a rule. So if ye dinnae mind...*'

'Fuckity fuck.'

'*Some respect, hen. We're in a graveyard, ye ken. Kekwick wis a marvellous companion. I couldnae huv asked for a finer man for the job.*'

'I don't believe this.'

'*Ye better believe it... and get used tae it. Let's just say you're ma latest travelling companion.*' Doesn't this man realise I'm a solo operator?

'*You've been talkin tae me in yer heid, puttin words in ma mooth, decidin what I wis thinkin. I thocht it wis time I stuck in ma ane tuppence-worth.*'

'This is *my* expedition.'

What am I letting myself in for? Hope I'm not going to regret this.

FOUR

TREE HUGGING

'No shower, no toilet, no facilities at all.' The guy at the homestead glances over at Bad-Ass Betsie, who is clearly minus any of those niceties.

'Perfect,' I say, and the last available spot in 'Bill's Paddock' is mine. Fellow campers are further along the creek bed, hidden beyond the gum trees.

Alpana is a working sheep station that has broadened into tourism to make a decent living, like many in the region. It's at the northern end of the Heysen Trail, named after the artist who described the Flinders Ranges as 'the bones of nature'. I feel I'm walking through one of his watercolours.

I used to think Heysen had a cavalier approach with the palette as he painted his beloved Aussie gum trees. How wrong I was. Bark hangs from the trees in peeling sunburnt strips, exposing patches of golden yellow, burnt orange and amber. Trunks lining the dry creek bed are embossed in 3D patterns. Some have the wavy rainbow patterns of an oil slick. Others are as smooth as satin. Tree hugging could be my next hobby.

I wasn't always complimentary about the Aussie bush, all those spindly branches lolling to the ground instead of pointing to the sky. Frankly, I found the bush untidy, a bit dusty, in need of a good clean-up. I mocked the gum tree's skinny leaves, flopping downwards and providing next to no shade at all in an environment where shade would come in very handy. Rather selfish, I thought, until I was warned that the gum tree is not a good choice to camp under anyway. Widow makers, they're called, for good reason, as they have a nasty habit of dropping very long and large limbs even when they're healthy.

As I fossick in the creek bed for wood for the campfire, hidden critters scuffle under the leaves, twigs and bark, letting me know I'm disturbing their homes among the gum trees. A better me would have brought my own firewood.

I upend a sheet of metal, making a shoogly but serviceable windbreak. A professional touch if I say so myself. 'Take that, Mr Stuart.' If there was one comment about McDouall Stuart that everyone agreed upon (apart from, you know, that he liked a tipple) it was his excellent bush skills. There's a story that McDouall Stuart loved camping so much that, when he returned to Adelaide, he sometimes camped in the backyard of friends' homes rather than stay in the house. My billy, filled with water, is building up to a lively jig for my two-minute noodles. That Girl Guide campfire badge is mine.

Time for my tent's virginal outing. I've only just straightened my spine from sleeping in the confines of Bad-Ass Betsie's rear so, beasties or no beasties, I'll take my chances on the ground. I never should have tried to change her, putting that floating floor in the back for a makeshift bedroom, making her into something she's not.

Spread out tarp and pop up the mozzie dome. Unfold canvas swag and help the mattress unfurl, with a bit of a shove. Pull sleeping bag from its sack, then layer each component inside. Quite a rigmarole. This will become a nightly routine as I learn to love my airy mozzie dome, not private but great for looking at the stars. The final touch is placing sandals within reach for the midnight toilet trek, torch (same), book (to help me fall asleep) and mobile phone (to tell the time). That book never gets opened and telling the time becomes redundant.

I sniff the eucalyptus in the air, sniff the smoky campfire, sniff the cheeky drop of red I've poured from the wine cask. Exploring is a thirsty occupation. I'm in sniff heaven.

Sipping my nightcap, I stare into the still-glowing embers of my campfire. It takes me back to the coal fires of my childhood, when I'd make up stories about castles and dungeons, having to start all over again when the coals fell in on each other. I'd sit so close I'd give myself tartan legs. I have a theory that the more you poke a campfire the less knowledgeable you are about keeping it alight, and I'm doing a fair bit of poking. Wonder what McDouall Stuart thought about when he was sitting around the campfire in the early days. Is it my imagination, or can I smell pipe tobacco?

'Stop footerin' aboot wi that stick in the fire, for God's sake.'

Martinet was one of the descriptions given to McDouall Stuart. Some men on his expeditions walked off the job because of his fierce standards. He had no time for people who shirked work or complained about conditions.

'I'll hae a drop o' what yer havin' if ye dinnae mind, hen.'

'Thought you didn't drink on your expeditions?'

'Ah didnae. I was in charge and had to set a guid example. But this isnae my expedition, is it? This time I'm a tourist, along for the ride.'

McDouall Stuart's break from surveying in the Flinders Ranges into the world of exploring came when Finke decided to bankroll his first expedition as the leader, which he ran like the ex-army man he was. Precision plus. Every morning, each saddle bag for the horses, labelled and laid out in strict order, was carefully weighed and packed. Why do I feel he's rolling his eyes behind my back? I find myself poking that fire again. Is he checking out my haphazard packing?

I need to assert some authority here. At some stage, I might suggest he stops talking about my swearing and I'll ignore his smoking.

As we gaze at the sky, the Southern Cross appears in the Milky Way. The constellation is visible anywhere in Australia, just about all the time, and they say you can use it to see where north is, but I've yet to work that out. In the Northern Hemisphere, it's the Big Dipper's outer stars that point the way. I never worked that out either.

It could've been the red wine. Or gazing at the starry, starry sky that circles the Earth, but we did haver endlessly about the universe as we sat staring into the campfire, talking about Scotland and homesickness yet wanting to belong and to feel settled. He called Australia 'my adopted country' and I said this was the place I now called home.

When I emigrated in 1975, it was a heady time in politics, near the end of the Whitlam era nationally and in the middle of the Dunstan decade in South Australia. I found myself at the polling booth almost the minute I stepped on Aussie soil, four times in five years, thanks to being British.

I even got a say in changing the national anthem from 'God Save the Queen'. The winner was 'Advance Australia Fair' but I couldn't vote for a song that has the word 'girt' in it. 'Waltzing Matilda', about a jolly swagman shoving a jumbuck in his tucker bag was tempting. It came second. However, I voted for 'Song of Australia', as did the majority in South Australia, possibly because it was written by a South Australian. Hardly anyone voted for it in the other states and territories—possibly because it was written by a South Australian.

'I've heard that song, it wis aroond in my time.'

'It was a bit twee, I have to say.'

In McDouall Stuart's day, there was no parliament so no voting. Everyone stepping off the boats was a new chum, the Aussie accent was just developing, the term New Australian wasn't yet in use. Australia likes to think it is a classless society but that's wishful thinking. It's never been so. Early terms to create a class divide were Exclusives (free settlers) and Emancipists (ex-convicts). When South Australia got its first parliament in 1857, one of its first steps was to deal with convicts, who were sent packing. Then the law was ramped up to deal with ex-cons if they dared step foot in South Australia within three years of serving their time.

Seven of the first eight Europeans hanged in South Australia were either escaped convicts or former convicts. A local newspaper report on the execution of two men charged with stealing £5 and discharging a firearm stated, '...it is absolutely necessary that the hardened villains who escape from punishment in the neighbouring colonies be taught that they have no triflers to deal with here.'

In the 1970s, one classification to denote superiority was still how many generations back your family could go since the First Fleet arrived. This was the first-class ladder reserved for true-blues and at the top of the ladder were the descendants of those Exclusives, perhaps seeing themselves as equivalent to the British aristocracy. To have a convict background was a dirty little secret. It's now a badge of honour. I bet Geoff back in Burra isn't happy.

Next was the migrant ladder, with Brits on the top rung, the prize for coming from the 'mother country.' Propaganda that played on the ties of tradition and connections with Britain was seen as making Australia an attractive destination, but that didn't attract me. An ancient culture that was still living, a multicultural society with all its differences. That's what appealed. Something different. I didn't think my own differences would matter.

I was told I wasn't really a migrant. Maybe I could move onto that first-class ladder? But there was a price to pay, an expectation to 'fit in' and could I do something about the way I spoke. I didn't consider myself a Pom, especially as the word was often linked with 'whingeing Pom' and 'Pommy bastard'. I wasn't a New Australian either—a label given to non-English speakers. Many first-generation Australians lost their accents as quickly as they could. To fit in. Some sound more Aussie than Aussies. I still can't say 'g'day' but sneaked in my Australian citizenship ten years after arriving, even though to this day I'm not keen on Vegemite.

Aboriginal Australians had no ladder. They weren't counted on the Census until the mid-1960s. From the belief that they were a dying race, so not worth bothering about? Many Aboriginal children in urban areas were growing up being told their grandparents were immigrants. Once again, a way to fit in. Why I didn't have Aboriginal neighbours made me curious at the time. Clearly not curious enough.

Perhaps Aussies were battling their own identity crises, being

Australian citizens and British subjects at the same time. This lasted until 1984.

What does un-Australian mean anyway, FFS?

It was Don Dunstan, premier of South Australia in the 1960s and 1970s, who took notice of those unfulfilled Letters Patent signed by King William back in the 1830s. He used the founding document to introduce a form of land rights for some traditional owners.

'Letters Patent? Never heard o' them.'

'Me neither.'

'I wis too busy surveyin.' That's true. Settlers had paid their money and were waiting for their land to be pegged out. *'Never dawned on me the land wisnae there for the takin.'*

I wasn't about to tell McDouall Stuart that a fellow migrant from his hometown of Dysart, land agent Robert Cock, was all too aware of the Letters Patent and the recognition of Aboriginal people's occupation rights. It's recorded he felt morally obliged to pay interest on one-fifth of the purchase price of land he bought in Adelaide, insisting it was not a donation but 'a just claim that the natives of this district have on me as an occupier of those lands'.

'I can't talk, Mr Stuart,' I admit. 'Never entered my mind either to consider who owned the land in the first place.' And so, when the legislation, begun by Labor Premier Dunstan and finalised by Liberal Premier David Tonkin in the early 1980s, gave traditional owners land rights to an area in the far north-west of South Australia, well, I was too busy checking out a piece of turf of my own to notice.

Words like dispossession, invasion and frontier wars would sink in many years later.

Terra nullius. Who would have thought that legal snippet would be tripping off our tongues one day. There's debate about who coined it first, and its meaning eludes many, but what's important is that it's been the excuse to deem that Australia belonged to no one before Cook landed. And why First Nations people had no legitimate claim to the land where they had lived for tens of thousands of years. That view was finally overturned in the Mabo decision of 1992 and the *Native Title Act* followed in 1993.

The Act doesn't give First Nations peoples the right to claim ownership of their land; their rights are 'traditional', generally to hunt, fish and camp, just so long as no other group—for example, a pastoral company or a mining magnate—is using it. Before any rights are given, however, they must be able to prove a continuous connection with their land. Not an easy ask, considering settler-colonisation means they were shunted off their land, pushed onto missions, denied access. The legal process often takes decades.

The Adnyamathanha, a term meaning 'rock people', is made up of different language groups. And, within that, different dialect groups. According to the Adnyamathanha Traditional Lands Association (ATLA), they include Kuyani, Wailpi, Yadliawarda and Pirlatpa peoples. Today, their Native Title rights and interests are recognised over 41,000 square kilometres of land in and around the Flinders Ranges and, as traditional owners, they are involved in the management of Ikara-Flinders Ranges National Park, a camping, glamping destination for bushwalkers, artists, photographers and folk like Bad-Ass Betsie and me. They are one of the relatively fortunate groups.

Many other groups are still battling it out in the Native Title courts.

I slip into my swag, watch more stars fill the sky, and find myself looking back to 1975 and the day we got off the train in Adelaide. The awkward hellos on the railway platform with the Scottish family who'd agreed to sponsor us—strangers until that moment. 'Who do you think is better—Skyhooks or Sherbet?' was the pressing question I got from the eldest daughter, all of sixteen, so much younger than me. I was twenty-two, a married woman of nearly two years and a global traveller. A grown-up.

Crazy, strange... fragile, low. I became the lines of the lyrics of Skyhook's hit song 'Living in the 70s'. Then came Sherbet's 'Life', released a few months after we landed and I was *living my life, doing things my way, feeling free.*

Good job, we used to say, good job that you have to pay back the cost of the journey if you leave Australia within the first two years. It was a see-saw. One day, 'this wouldn't have happened if we'd stayed back in Scotland' (*crazy, strange*); the next, 'this is the life' (*feeling free*).

No coal fires to soot up the buildings. Under my feet, the pavements were free of litter; above was a sky so clear and blue it made me dizzy (*feeling free*).

The tyranny of distance was never so sore for us as when we were having a baby. Choosing the five-week boat trip to Australia over flying out—to make us realise we were going to the other side of the world—had done better than we intended, maybe too well, making it clear how far away we were from family and friends. We sent a letter home as soon as we knew I was pregnant (no one had a telephone), and we received the card congratulating us on our good news as I lay in a hospital bed having had a miscarriage (*fragile, low*).

I thought of another connection with McDouall Stuart. He still had to do surveying work for Chambers and Finke on his early expeditions. I'm the same. I'm funding this trip with a writing gig in Alice Springs at a conference for remote area nurses. That means I'm travelling through the centre a bit later than I'd like. McDouall Stuart, too, didn't have a choice on the timing of his trips.

There's a distant laugh of a kookaburra, which never fails to make me smile, and faint squeaks and mutterings drift across from the creek bed. Everyone is settling down for the night. Outback silence.

'This is the life,' I whisper to myself. What a difference a day makes. I'm going to watch those stars sparkle all night long.

A bird battle over perching rights wakes me at dawn. Two wee spyugs have decided to evict a big black corbie from a skinny-ma-linky stick of a tree that's not worth fighting over. Where did those Scots words come from? Haven't used them in a long time.

That tree is the equivalent of the worst residence in the best street, but the crow's tormentors are determined. He flaps off and the little naggers congratulate themselves. Then the crow is back for another round. Not sure who to admire more: the bickering little sparrows or the bloody-minded crow. I take lots of photos and videos. I'm no David Attenborough. I later learn this is common practice. They might not even be sparrows. Or crows.

I nudge the fire embers back into life as the rest of the local bird community sings and whistles into the dawn. The billy's back in place for a morning cuppa and I think about the good old chat that McDouall Stuart and I had last night. It could be okay having him along for the ride.

'What do you think Betsie?' I look over at her. 'Not as if he's going to take up much room, is he?'

I start a mental list of resolutions as I pack up.

- Keep Bad-Ass Betsie tidy. I promise myself I'll stuff that sleeping bag in its sack every single day.
- Keep myself tidy. Buy a pumice stone is top of that list. My heels are already looking as cracked as the creek beds around here.
- Buy a star chart of the southern skies. I'm going to have lots of free time to educate myself about all sorts of things—stars, geology, trees, birds. A fine idea.
- Check the satellite phone is charged and work out how to use it. Not doing much good sitting in its yellow security box.

A file of bushwalkers at the tail end of the Heysen Trail cross the creek bed and disappear over the hill as I drive off.

I hear a muttering *'Well hen, let's see how that list survives.'*

FIVE

FIRST ASTERISK ON THE MAP

An inflatable kiddies' paddling pool takes up prime position in the garden of a clean-cut prefabricated house. A standard home and a typical summer scene in the suburbs, except this vision of domesticity is at the end of a long, long drive past sun-crisped fields and soft-topped hills that roll on forever. This is Moolooloo, 500 kilometres from Adelaide.

Lulled by the car's air-conditioning, I'm shocked by the heat when I hop out; the red dirt seeping into my sandals burns my feet and the hot air grabs my throat. I'm walking on land McDouall Stuart would have walked over many times, first of all surveying it for the Chambers brothers, then later preparing for his first expedition as leader.

My, how this pastoral property would've changed since the mid-1800s. Those massive sheds shimmering in the sun for a start. The dirt roads and paths wouldn't have existed. The heat and the flies; nothing's changed there.

Layers of early-settler life are embedded in patched-up buildings, one with a roof of buckled corrugated-iron sheets revealing splintered wooden beams and clumps of dried-up thatch below. Rusting metal is scattered all around—machinery long past its use-by date poking through wild lavender.

Loud barks lead me around the corner, where I find Keith Slade and his dad, Keith Slade Senior, on the back of a truck, grappling with a serious piece of farm equipment. It's the height of sheep-shearing season and Keith the Younger had warned me when we arranged my visit that he'd be too busy for a chat. He stops long enough to tell me I'm welcome to camp by the creek bed or set up among the bales in the woolshed. My choice. You must go up Sunset Boulevard for a great view of the whole property, he says, pointing to a path leading up a hill. Father and son turn back to their work and a bad-tempered cloud of flies rises from the sweat patches on the backs of their shirts.

I check out potential camping spots along the length of the dry creek bed but give up on that idea. Haven't those flies got some sheep to pester? Sunset Boulevard is narrow, covered in slippery shale with prickly bushes on either side, and as I nudge Bad-Ass Betsie onward, I hope they're more tickly than scratchy. I'm well aware I'll have to turn her around and go back down the way I came.

The scene from the top is pure Hollywood wild-west movie and, as if on cue, a cloud of dust appears in the distance. It's a jeep, so far below it moves past in silence, following a faint track leading to a range of blue hills beyond. A thirty-three-point turn gets Bad-Ass Betsie facing back down the hill. I can almost feel her cringe with embarrassment. She's more than capable of a swifter turnaround but I'm not. As for McDouall Stuart, he'd better say nothing.

The woolshed isn't the one in McDouall Stuart's day, but I reckon there could be a few fence posts in the yard outside dating back that far. Lounging on my mattress inside at the end of the day, I'm surrounded by piles of fleeces, overflowing bales ready for the wool press and tightly-packed cubes lining the walls, the smell of lanolin all around.

Already I'm slacking off with my diary. I'd every intention of noting daily distance travelled, places passed, changing landscape, the weather—just as McDouall Stuart noted wind directions, meticulously calculated map coordinates and notable landmarks.

Chores like making a cup of tea swallow up the hours. Attach the bits and pieces for the gas, evict creepy crawlies from the billy and ignore the wriggling goings-on in the bottom of the water container. Where are the matches, the teabags? My stuff in the back of the van, well-packed when I left Adelaide, is becoming less so with each successive night. I've given up rolling and packing the sleeping bag and the self-inflating mattress. McDouall Stuart knows me better than I do. The food box, utensil box and toolbox are now interchangeable, with packets of biscuits sharing space with the hammer, and the cords for all the gizmos wrapping themselves around the bandages and fly spray. That's where I find the tea bags.

McDouall Stuart's first expedition as leader was a modest affair with two companions, a handful of horses and enough flour, sugar, tea, salt, tobacco and jerked meat to last four weeks. He was instructed to check out land for pastoral options and also mining options, preferably with a streak of gold. But his journal entries reveal he harboured another dream: to head for the centre.

He was also fair taken with the idea of locating Wingillpin, a legendary Aboriginal paradise, north of the Flinders Ranges, that he'd heard about. He never did find Wingillpin, nor the lucrative lodes of gold that Finke had hoped for. However, he did find 'as fine a creek of water as I have seen in the colony'. Were it near Adelaide, he wrote, 'it would take its place as one of the South Australian rivers, and not the least by a long way'. He called it Chambers Creek.

His allotted four weeks were well and truly up by the time he reached Chambers Creek but, rather than take the shortest way back here to Moolooloo, McDouall Stuart headed south towards the Great Australian Bight, mapping 103,600 sq kms of unchartered territory. When I read his journal it's hard to comprehend why he made that choice.

If I thought the shale up Sunset Boulevard was tough going in sturdy walking boots, McDouall Stuart was writing about 'large fragments of flinty stone' and 'stiff tenacious clay' sticking to the horses' hooves, making them lame. He felt sorry for his beloved horses, but he was critical too. He had a go at them in his journal for being fussy, writing: 'Some of the horses would not drink the water, and others drank very little: they will be glad to drink far worse than this before they come back, or I am much mistaken.' Oh my, did I hear that sort of sentiment growing up. Must be a Scottish thing.

He ran out of horseshoes, did nothing but complain about the lack of water and then too much water and his rations went down to the bare minimum.

'What were you thinking?'

'*I wanted tae see what wis there.*' He shrugged. '*I didnae intend tae go that far south but then I saw an important feature I couldnae leave unexamined. I called it Mount Finke.*'

'As you would, him being your benefactor. And?'

'*Well, I got on one o' the lower spurs to see what wis afore me. The prospect wis gloomy in the extreme. Nothing met the eye save a dense scrub as black and dismal as midnight. Wisnae ony better the next day frae the top. A fearful country. The land wis a bit hit and miss I huvtae say.'*

'Tell me about the mirages.'

'At one point, the mirage was so bright, and went on and on, ye'd think the whole country wis underwater. But I kent better.

'Mirages can be so powerful. Some folk think they're only aboot water but a hump in the landscape looks like a hill, or even a mountain, and wee bushes appear like muckle big gum trees where there's nae timber at a'.'

One of two men on this first expedition, an Aboriginal tracker, left them to it somewhere on the way. He'd had enough. The other, a man called Forster, stuck it out. Only just. McDouall Stuart and Forster reached the coastline at Streaky Bay in a pitiful state. They were welcomed at the sheep station there and had supper after three days of enforced fasting, which made them violently sick. Then they set off back to the Flinders Ranges. The four-week expedition had turned into four months.

'I liked your story about pigface in that first journal. You started off saying it was good for the horses.'

'I ken. The horses liked the stuff. There's a lot o' moisture in them, first-rate for thirsty horses.'

'Then you ended up eating them yourself. Pigface is now on the menu in some swish restaurants around the country, so you were a trendsetter.'

'I could gie them ma recipe eh? We boiled the pigface in their juice. Huvtae say, to a hungry man they were very palatable. Had they been boiled in fresh water, they would've made a good vegetable. One day we added young sow thistles, boiled, and we could have enjoyed more o' them.'

'Not so sure about your hopping mice recipe, Mr Stuart. Even though you call them 'elegant little animals' in your journal.'

'*Pot calling the kettle black, eh? I've watched you cook yer two-minute noodles.*'

Robert Bruce, the Arkaba Station manager, had a few words to say about McDouall Stuart's appearance on finally arriving back on his Flinders Ranges property. 'A sharp voice with a Scotch accent accosted me from the fence,' Bruce wrote in his Reminiscences. 'I turned to see a pallid, pasty-looking face, crossed by a heavy moustache, and roofed in with a dirty cabbage-tree hat, peering between the rails.'

'*That Bruce could tell a good tale for sure. "What have you been doing with yourself?" he asked me. "Your voice is all that is left of you." I soon told him. "Well, if you'd been living for weeks on mice and pigfaces, as I have, there widnae be so much left o' you either, I'm thinking. I've had a terrible rough trip".*'

Robert Bruce was to write later about McDouall Stuart: 'He had the pluck of a giant in his puny frame, coupled with a prudence and good judgement that eminently fitted him for the leadership of men.'

One of the most daring exploits performed in the colony. That's how the Governor of South Australia, Sir Richard MacDonnell, described McDouall Stuart's first expedition on his return to Adelaide.

McDouall Stuart thought so too, and worth a piece of that land at Chambers Creek as a reward for his efforts. This is the first record of him being in the slightest bit interested in having a patch of Australia for himself. The request did not go down well with the fledgling South Australian government.

The Royal Geographical Society of London, however, gave him a gold watch for his efforts. So he did get gold of sorts after all.

'That was nice, getting the gold watch. I don't buy what you wrote in your journal though, suggesting you weren't fussed about not getting to the centre this time around.

'Let me read it to you, it deserves to be read aloud.' I clear my throat and speak in my poshest pan-loaf accent: 'No doubt I have opened up a great field for discovery in that direction. At any rate, if not for myself, I have cleared a tract for others to follow and extend.'

'*Aye. So?*'

'Who were you trying to kid? You're like a Scotty dog or an Aussie camp dog. You were never going to let go of your dream of reaching the centre.'

SIX

ANCIENT TRADE ROUTES

I'm on the Outback Highway stretching from Hawker to Marree, leading me to the start of the Oodnadatta Track. The road, now bituminised, follows an ancient trans-national network of trading routes crisscrossing the entire continent, used by Australia's First Peoples. Trading routes are all over the world and this is one of the oldest.

Spear shafts, ornaments, spinifex gum, shells and bush tobacco, all found far from their original locations. Native tobacco (pituri) grown in south-west Queensland and Central Australia, a mixture of leaves and wood ash traditionally chewed as a stimulant, enjoyed a roaring trade. It would be packed in special net bags for transporting hundreds of kilometres from the source. Pearl shell from the north-west of Western Australia travelled further perhaps than any other object, as far as the Great Australian Bight in South Australia. Baler shells from Cape York in Queensland, chipped, ground and perforated to make oval ornaments up to ten centimetres long, have been found at Kati Thanda Lake Eyre and in the Flinders Ranges. But the material believed to be traded for the longest time and one of the most desirable is ochre, used for body art, cave paintings and ceremonies.

I'm looking over a cliff face at Lyndhurst, lined with seams of red, orange, yellow and white ochre. Despite the nearby placard for tourists, however, this is not the ochre that brought traders from afar, as the quality was considered poor. The site looks spectacular all the same. Not far from here, near Parachilna, is the site where the ancestors of today's Adnyamathanha people extracted a red ochre considered highly significant. There are numerous accounts that this was where Aboriginal peoples from all over Australia headed, bypassing other more accessible sources.

To prepare ochre for carrying home, often hundreds of kilometres away across Australia, the ochre was mixed with water from the nearest waterhole or spring to form a paste which was then formed into a block with a slight hollow so it could be carried on the head.

The simplistic view by early Europeans that these journeys over the country were primarily about economic transactions has been knocked on the head. The late Professor John Mulvaney wrote extensively about the gradual realisation that these trips held deep spiritual and ceremonial value. The purpose of these well-trodden paths was less about supply and demand and more about social and ritual needs. The men sent on these trips, carrying message sticks and gifts, would socialise, taking the opportunity to hold ceremonies and teach each other different dances to pass on stories.

They would no doubt exchange ideas, perhaps give travel tips and pass on news. A hot topic since the start of colonisation would have been about the invaders, be they explorers or adventurers, squatters or settlers. A common theme in oral stories of first encounters is that these interlopers on horseback were strange minotaur-like beasts with human heads, or ghosts of ancestors who'd been skinned, which would explain their whiteness.

'I can tell you aboot one encounter I had near here.' McDouall Stuart has my full attention.

'A lot of the time you used tae could tell they wirnae far away. We'd see smoke, but when we got close...naebody wid be there. Figured they'd seen us and run aff.

'This one time, I came suddenly upon a native who wis huntin in the sand hills. What he imagined I wis I dinnae ken, but when he turned aroond and saw me, I never beheld a finer picture of astonishment. Riveted to the spot, he wis, his mooth wide open, and his eyes starin.

'I sent oor black forward to speak tae him but forgot to tell him tae dismount.

'Well, you should've seen him. He threw down his waddies, and jumped up into a mulga bush, as high as he could. I swear he jumped so high, one foot was three feet from the ground, and the other, two feet higher. I expected any moment tae see the bush break wi' his weight. He kept wavin us aff wi' his hand when we moved towards him.'

Before this trip, I'd noted the comments by the late anthropologist and linguist Ted Strehlow: 'It is greatly to the credit of Stuart that he was an infinitely more humane man than many of his white contemporaries ever aspired to be.'

I also read views suggesting engagements with explorers were distinguishable from encounters with pastoralists and prospectors. That made sense to me. Explorers would be passing through the country as quickly as possible, not hanging around for long. McDouall Stuart was certainly time-poor. He would read Aboriginal signs and the landscape in the search for water, but avoided direct and potentially risky encounters. I was happy to believe that McDouall Stuart was consciously non-confrontational.

It figures that drovers taking cattle to the pastoral stations spreading up the country would take the path of least environmental resistance, around mountains, deliberately following helpful water sources, without knowing they were using paths well-trodden already by First Nations peoples. Ancient trade routes meet colonial stock routes.

First Nations peoples soon realised that encounters along these routes were not with ghosts; and that these people were rudely crossing over clan boundaries without a by-your-leave, letting their four-legged beasts drink the precious water and trample the waterholes. As pastoralists made it clear they had plans to stay, what began as harmless encounters, swapping trinkets and providing directions to water, led to confusion and then turned into conflict.

In 1860 a Chief Protector was appointed to watch over the interests of Aboriginal people in South Australia to quell the violence on the frontiers, to reduce devastation by disease and to provide rations as their natural food sources diminished. South Australia is where the phrase 'smooth the dying pillow' was first used.

At Federation in 1901, the consensus was that Aboriginal people were a doomed race. The national government couldn't see any reason to give Aboriginal people the right to vote or to acknowledge them in the Constitution.

Keen to learn about Australian history when I arrived in 1975, I joined an evening class to find only one session that mentioned Aboriginal people. I read numerous history books that began with

Captain Cook, no mention of Aboriginal culture. Others had chapters on pioneer settlers, never mentioning whose land they were settling on; and chapters on Gallipoli and returned soldiers, but no mention that Aboriginal soldiers were among them, and were denied the same compensation of land for their war efforts. Many Australian friends later told me they learned next to nothing about Aboriginal history and culture at school.

The blank spaces in history, coupled with false stereotypes—such as those pioneer memorials still present in towns around the country of brave white settlers and savage natives—well, is it any wonder studies and surveys to this day show a huge percentage of Australians have a negative bias against First Nations peoples?

Events that occurred only a few years before I arrived were not on my radar. The Freedom Ride in 1965, inspired by the Freedom Rides in the United States, highlighted segregation and the appalling living conditions of Aboriginal people in New South Wales. The resounding 'yes' vote in the 1967 Referendum changed the Constitution so that Aboriginal and Torres Strait Island people were counted for the first time in the Australian Census. In 1971, the Aboriginal flag, designed by artist Harold Thomas, a Luritja man from Central Australia, was first flown on National Aboriginal Day at Victoria Square in Adelaide. The next year the Aboriginal Tent Embassy was set up on the lawns opposite Parliament House in Canberra to protest the Federal government's handling of land rights. The Tent Embassy is still there, now embracing issues such as sovereignty and self-determination.

Every year there are debates about the appropriate date for Australia Day, you'd think it was a fairly recent consideration. But no. Only recently I also learned that on 26 January 1938, while many

Australians celebrated the 150th anniversary of the landing of the First Fleet, a group of Aboriginal men and women gathered in Sydney for the first Day of Mourning. Those celebrating held a parade and a sailing regatta; those who mourned were forced to march in silent protest (and go through a back entrance into the hall where they had planned a meeting).

So much for First Nations peoples being compliant, so much for not being interested in land rights and land use, so much for dying out.

SEVEN

TOWNSHIP OF BROKEN DREAMS

The streets of Farina are wide open and welcoming, sweeping past generous portions of pegged-out blocks of land, nearly 500 of them. The image of a town planned for prosperity.

I'm a few kilometres south of what's now the start of the Oodnadatta Track, visiting the first of three station properties in the area, each with a different vision to address past environmental mistakes in the region.

McDouall Stuart is remembered as the first explorer to pinpoint waterholes at this spot, originally called The Gums, then Government Gums. Years later, someone had the bright idea to call it Farina, Latin for flour, setting up the town for ridicule.

Surveyed in 1878, Farina was earmarked as a central point for a wheat belt fancifully filling the map right up to here and beyond. The 'granary of the north' it was to be. Homes were built, mostly a combination of wood, canvas and tin. The fancier ones had earthen floors, covered in linoleum. Public buildings were built of local stone, the mortar often minus the lime needed for bonding. In no time, Farina had an underground bakery, a bank, two breweries, five blacksmiths and a brothel.

The first wheat crop was a success but by the time the rail line arrived in 1882, there was never to be another decent harvest. Farina was to become a ghost town, a permanent reminder of the era when people believed in the myth that 'rain would follow the plough'. Not a sheaf of wheat is in sight today, but the flies are thriving. I'm grateful for the flynet over my hat.

In the campground, I have my own personal fire pit and quite a bit of real estate, so I park Bad-Ass Betsie to one side and pop up the tent on the other, positioning the chair to take advantage of the view into the distance. I look over to Bad-Ass Betsie. She looks settled too.

I do appreciate that the toilet, at the edge of the paddock, is a sweet-smelling, flushing loo, with a little basin attached to the outside wall, boasting a bar of soap. Not the standard Outback long-drop. I'm easily pleased. The shower has a water heating contraption (it's called a donkey boiler, don't ask me why), thankfully unnecessary with today's temperatures. With the resident family of emus watching me, I would have been a bit self-conscious trying out my skills at stoking the fire.

A rumour is circulating that you can get mobile reception at the top of a nearby hill so here I am, joining a dozen other campers waving their phones in the air for a signal. It's a flynet fashion show. Pastel pink and blue on a couple of kids, a snazzy rainbow one, another I swear is made of bridal tulle, each one making my army-fatigue version look dull. Even the men sport black, grey and green varieties. Gone are the macho days when no self-respecting bloke would be seen dead wearing one. Standing on top of a bench and reaching skyward fails to give me one bar.

I trudge along a belt of hot red sand, past a piece of farm machinery once pulled by a camel and come across a fenced-off area, an open-air museum of ironmongery where a monster cabinet displays branding irons and buggy steps, buckles, bolts, bits, braces and bobs. I know this because each one is labelled and categorised. There are spanners, tongs, punches, chisels and spikes, copper rivets to repair leather, a crank from a grindstone and a tip for a plough: rusty relics of life here last century revolving around farming, mining and rail transport—each dream gradually shattered.

By the time I reach the outskirts of the town, sweat is trickling down my face. The sunglasses have no chance, slipping off the end of my nose and into the flynet. Squinting in the sunlight, I read a sign saying beware of tumbling walls. Gaping holes, once doorways, lead into roofless rooms with not a smidgen of shade. Even the local wildflowers are keeping their distance. Gaps that once were windows gawp out over the barren gibber plain where farmers once ploughed the fields to sow their seeds of barley and wheat.

A sign at the Exchange Hotel tells me mutton was always on the menu, serviettes were placed each day on the tables and patrons were expected to dress for breakfast. This may be the outback, but decorum was expected in the dining room at all times. That wasn't the case in the bar and locals were relieved at the end of many a weekend when the shearers staggered back on the train after partying until their pay cheques were spent.

The hotel closed in 1937, the cemetery was last used in 1960, and the railway station closed in 1980 when the entire train track, which now stretched to Alice Springs, was moved further west.

Most of the houses made of corrugated iron sheets blew away and the stone buildings crumbled. By 2000, Farina was marked on maps as 'ruins.'

Another sign announces the annual visit from the Farina Restoration Group. Farina has the last laugh. Group members from all over Australia arrive in their caravans and campers in the cooler winter months of June and July to stabilise the buildings, funding the restoration program by selling bread, pies and cakes baked in the fully restored underground bakery. Over those two months, hundreds of travellers a day take the sizeable detour along the Outback Way to taste the goods.

Farina has become an apt name for the township.

Conservation management is the path taken at Witchelina, a nearby run-down pastoral property taken over by the Nature Foundation after the drought of 2009 forced the leaseholders to sell.

Much of the foundation's work revolves around dealing with pest animals and plants introduced to the land since colonisation. All domestic livestock and most of the feral goats have been removed, and they are also tackling degradation caused by rabbits. Programs are underway to reduce foxes and cats, and another one deals with buffel grass, which is recognised as one of the greatest pest threats to the arid rangelands of South Australia.

It's easy for us to shake our heads at the damage caused by first settlers who blundered over the land, treating it as they would back in 'the mother country', creating a landscape of parched lands suffering from plagues of introduced pests and dust storms. We're still making environmental mistakes though and studies show Australia is the fourth-worst country in the world for species extinction and land clearance figures are still high.

The foundation works with the Adnyamathanha, Kuyani and Arabana traditional owners of the land, not only on conservation measures but also to increase cultural understanding. Witchelina hosts several Kids on Country camps each year, a program that builds employability skills for young Aboriginal people and sparks their interest in conservation and land management.

The third property I'm about to visit is the expanding Wilpoorinna Station where Gordon and Lyn Litchfield, both ardent environmentalists, began changing their practices many years ago, starting with ditching their flocks of Merino sheep—almost a sacrilegious move in Australia.

Gordon's family were among the first settlers who came to this region as the outback opened up in the mid-1800s, but Gordon, a legendary cattleman in these parts, is no traditionalist.

Along with switching from wool to meat production, the family recognised that increasing their land mass was necessary. Wilpoorinna Station, together with adjoining Mundowdna, covers 2590 sq kms, plus another acquisition, Mount Lyndhurst Station, takes their holding to 6070 sq kms. It's a lot of land, but I soon learn there are bigger stations further into the middle of Australia—some the size of a small European country.

'We must recognise that old approaches have not worked. In fact, they have worked against the health of the country,' explained Gordon on the phone when I arranged my visit. 'We've taken on a new approach, looking at how to work with the land, paying attention to the land, rather than trying to be the master of the land,' he said. 'We've been working towards this for a long time. People tell me we're lucky. And I say "Yes. And, the harder we work, the luckier we get".'

EIGHT

A TOUGH LITTLE BUGGER

'Turn right at the big rock' are the scant instructions for Wilpoorinna cattle and sheep station.

Long gone are the towering cliffs and deep craters, spectacular gorges and sheltered creeks of the Flinders Ranges. Replaced with the stony plains and deep-red soil of South Australia's arid lands. Not a blip on the horizon. I start to worry if I'll see any rocks. Of any size. Ever again.

Gordon, a fellow McDouall Stuart tragic, is keen to show me the monument he's built in McDouall Stuart's honour, two metres tall and weighing five tonnes. First, however, I have to find their property.

And then, there it is. A big rock points the way to Wilpoorinna homestead at the end of a 25-kilometre driveway.

Perched at the kitchen bench with a cuppa, Lyn and I are having a good blether—occasionally allowing each other a word in edgeways. Already we've exchanged stories of our children and grandchildren. Lyn has invited me to stay the night in the guest cottage, and I say I'll treat her and Gordon to dinner tonight at their local pub in Marree, fifty kilometres further up the Outback Way.

We agreed that 1975 was quite a year for us both, the year I emigrated to Australia from the other side of the world and the year Lyn took the 700-kilometre train ride from Adelaide to Marree, into a different world. She intended to complete a short stint as an outback nurse before heading to Africa to train as a midwife. Within twelve months, she met and married Gordon and she abandoned her plans to travel to Africa.

'Everything I've done in my life seems to have woven together in some way,' says Lyn, and I stop myself from telling her that McDouall Stuart and I were saying the very same thing a couple of nights ago around the campfire.

We go for a walk, joined by their wee dogs Charlie, Sophie, Pepper, Smokey and Flash, barking in delight as they circle Lyn's heels. 'Goanna.' Lyn points to a line on the path. 'It's mating season, so there's lots of tracks.' She points to a wavy line. 'And that's a snake. See the direction he's been pushing the earth... shows he was moving away from the cottage.' I'm glad to hear it. Tracks of an emu, a sleepy lizard, a crow...they all looked like scratches in the earth to me. Now I know better.

Lyn, a nurse at the Marree clinic for forty years before she retired, picks some wildflowers near the homestead. 'At night, the perfume is beautiful.'

I guess she's got time to smell them now. Before the family grew up and before retirement, Lyn's hectic routine in the outback, where the niceties of city life are beyond reach, included receiving huge cartons of books from School of the Air to help tutor the kids, and, in her role as a remote area nurse, working in tandem with the Royal Flying Doctor Service.

As we pass a paddock, I try to take photos of a mare with her foal, but all I get is her backside as she turns away. 'Horses are very sensitive, very intuitive,' says Lyn and I wonder if she senses that me and horses don't go together so well.

'There's an old bushman's saying that the outside of a horse is good for the inside of man,' says Lyn. She's studied equine-assisted therapy and believes in getting people out of the counsellor's office and onto a horse.

'Horse whispering. That's for the movies. It's more body language, the gentle pressure and release of your hand.

'Horses respond through the heart and desire more than through force. A horse can feel a fly on its back. You need to give them your full focus. You can't have random thoughts.' Is it too late for me to learn?

'They've got a herd language and if you can learn the herd language, you are developing horsemanship skills.'

I've only had one experience with horses, many years ago. A group of us hired some horses for a canter or whatever it is you do with horses and I was given the slowest one considering my inexperience. I think I pissed him off. I'd been told to knee him to make him start trotting. I hadn't been told how to make him stop. I pushed my knees against his sides, he picked up speed, caught up to the others, passed them at a dizzying pace and continued galloping, at one point going under a low-lying tree so I had to crouch like a jockey to avoid being whipped off by a branch. The snot and the tears were sliding past my ears when I heard the thundering of horse's hooves and I had a vision of my saviour grabbing me round the waist and pulling me to safety.

'Stop shouting and he'll stop galloping.' That shut me up. Somehow my saviour manoeuvred the horse uphill to slow it down and I forgave him. The man. Not the horse.

We're back at the kitchen bench and the phone rings constantly. I thought life in the outback would be quiet and laid-back. It has its moments, says Lyn who's telling me how she introduced the benefits of massage to a wary clientele in Marree during her time as the local nurse by calling it her 'horse treatment'.

Since childhood, Lyn has done hands-on treatments on her horses, massaging them for shin soreness and injuries. When Norm, a patient at Marree clinic with diabetes and peripheral nerve damage, was at the limit of his medication level, Lyn offered him her 'horse treatment'. Word spread and led to Lyn gaining massage qualifications and permission to offer this service at the clinic. She reckons this was partly thanks to Norm's glowing testimonial.

He wrote: 'When I came to Lyn, I couldn't feel my hands or my feet; I couldn't sleep for more than two hours; and I couldn't walk without the walking frame. Now I can walk all the way to the pub; I sleep all night; and I can feel Lyn's hands all the way to my heart.'

'Healing... it's the same for land as it is for people,' says Lyn. 'If someone comes into the clinic with a headache, you check them out, ask some questions, try to find out why they have a headache. You help them change the way they do things, improve their health management. It's the same with the land. If you see erosion, you do a bit of checking, work out what's causing the problem and then find solutions.'

Gordon opens the door and if I hadn't heard the ute pull in I'd swear he'd tied up his trusty steed outside. It's the stockman's hat that did it.

We're introduced, but there's a whirlwind of phone messages to deal with first. Gordon strokes his healthy handlebar moustache as he checks the wall calendar filling up with more commitments—a funeral to attend, a rodeo coming up, arrangements for sheep to be trucked—before turning his attention to me.

He unfolds and flattens a map over the dining table, slides off his hat and rubs his head as if he can't believe the stories he's telling me. Of rivers that run in the wrong direction, away from the coast, only to disappear in the middle of the continent. Of the now-you-see-it-now-you-don't salt lake that is Kati Thanda Lake Eyre. It's twenty times bigger than Scotland's Loch Ness when it's full but with a habit of disappearing for decades. Of water that's two million years old when it spouts out of the ground, not far from where we're standing.

'Here's where we are.' He stabs the map. 'Here's Marree. This is Lake Eyre. Lots of explorers came this way.' He's stabbing away. 'Lake. Hit. Lake. Hit. Then another fella comes along. Same thing. They kept hitting the edges. It was a lake like they'd never seen before,' says Gordon. 'For a long time, they reckoned you couldn't get through.' Explorer John Eyre died thinking the salt lake that bears his name was one massive impenetrable horseshoe of mud.

When explorers discovered that the horseshoe lake was several lakes with gaps leading north, they were relieved to find some springs in the parched land beyond. But it was McDouall Stuart who mapped a whole string of them, stepping stones to reliable water sources. He realised early on how important they were going to be for exploring north through some of Australia's most inhospitable desert country and into the interior of the continent.

Gordon outlines the route McDouall Stuart took, a route that the Arabana Nation had always known of and used, particularly in times of drought. The springs were the only permanent, potable water in the region that could be relied upon when surface water in claypans and rock holes dried up.

Archaeological digs along the Track have revealed grinding stones, flints and stone tools close to the mound springs. Research and oral history gathered from Arabana people also make it clear the springs were important for their day-to-day life and had deep cultural significance, making them key sites for significant events. It's no surprise the springs feature strongly in stories and songs of the Arabana, whose Native Title land stretches from Marree in the southeast, past Kati Thanda Lake Eyre and the Wabma Kadarbu Mound Springs Conservation Park, up to Oodnadatta in the north-west.

That's where I'm headed. But first I have a date with Stuart Man. And dinner at the Marree pub where I'm promised I'll meet another McDouall Stuart tragic.

We're on the northern edge of Wilpoorinna station, thirty minutes after leaving the homestead, and there stands Stuart Man, legs apart and arms outstretched, bracing himself against the wind that's whipping dry red dust around him. The vast, flat land stretches on and on, the huge Australian sky touching down on a horizon far off. 'No pastoral lands are as dry as we are here in this part of Australia,' says Gordon. I believe it.

I've persuaded Gordon to let me take his photo in front of his monument, but nothing will make him take off his stockman's hat, so it takes a while to get an angle that doesn't throw too much shadow over his face. Lyn is sitting in the car. Sensible.

Gordon first heard of McDouall Stuart as a youngster when a horseman rode through the family property in 1962. This traveller was retracing McDouall Stuart's final expedition a hundred years earlier. 'The stories he told about McDouall Stuart, that stuck in my mind.' Gordon reels off camping spots named in McDouall Stuart's journals but no longer used. Places where McDouall Stuart lost his horses, found his horses, watered his horses.

'It's one thing not knowing what he was letting himself in for the first time he came through. Going back up there, again and again...I loved hearing those stories,' Gordon tells me. 'Polly was McDouall Stuart's favourite horse and my Dad, he had a creamy mare called Polly. That stuck in my mind too.'

No wonder Gordon is a McDouall Stuart fan. For me, all it took was a magazine article to pique my interest.

'In 2012, I was thinking, what could I do to commemorate the 150th anniversary,' says Gordon. 'McDouall Stuart's journals talk of cairns.' He points to a cairn on a nearby hill. 'Then I saw the icon for the Canadian Winter Olympics called 'Five stones make a Man'. I thought of the quarry that's near here and it all came together.' With the help of some mates, a loader and a tip truck. The largest stone for Gordon's homage to McDouall Stuart weighs one-and-a-half tonnes.

We retreat to the car, escaping the dust and the flies. 'He took just what he could fit on the horses,' says Gordon. 'Unlike some of the others, who needed bullock wagons to carry their stuff.'

On his first expedition, McDouall Stuart didn't even have a tent. He took no scientific equipment and minimal navigation gear. In comparison, the gear that explorer Benjamin Babbage took on an expedition around Lake Torrens at the same time included six iron tanks capable of holding 5,000 litres of water and a camera on stilts.

Catering on McDouall Stuart's first trip was dried meat and flour. Babbage's food stocks included thirty-eight kilograms of German sausage and twenty kilograms of chocolate.

On the way to Marree, another thirty minutes away, the three of us try to remember more statistics. Burke and Wills, also in the outback at the same time as McDouall Stuart, had a similarly colourful list of provisions. Lyn recalls hearing a story about an explorer taking a hundred pairs of trousers. We decided that that man must have been somewhat fastidious.

'McDouall Stuart was a tough little bugger,' says Gordon in true outback understatement.

Not one for outback understatement is Phil Turner, publican at the Marree Hotel and a walking-talking promotional billboard for the town. He's my second McDouall Stuart tragic of the day.

'He's Australia's greatest inland explorer,' says Phil. 'Burke and Wills. What did they do for Australia? Who were they? Idiots. That's what.' He had more to say on the matter.

'People who come through here have heard of McDouall Stuart but have no idea what that man achieved. They don't know that so much of what's happened here is because of his explorations.

'McDouall Stuart is so underrated.'

Phil's contribution to lifting McDouall Stuart's presence in the area is a display in the pub's dining room focusing on the Overland Telegraph Line, which followed McDouall Stuart's tracks.

The Line, built in two years, is considered one of Australia's greatest engineering feats of the nineteenth century—a single strand of wire strung between 36,000 poles spaced eighty metres apart from one end of the continent to the other. This was the missing link to provide speedy communication, via Morse code, between Australia and the rest of the world—a vast improvement on the six weeks for letters and newspapers to arrive onboard a sailing ship or steamer.

Construction began less than ten years after McDouall Stuart's final expedition. His maps were so accurate, says Phil, that a construction worker on the line was overheard saying, 'I would believe that plucky Scotsman's instructions any day', or words to that effect.

McDouall Stuart, chatty for once, was in touch early on with Charles Todd, the man who was to plan and organise its construction, and he was careful to note in his journals the best places for river crossings and sources of water and timber for the telegraph poles.

Although the Overland Telegraph Line was regularly damaged by bushfires, floods and termites, it continued to be used until the 1970s. The repeater stations built along the way are now crumbling relics of the days of dots and dashes, so I won't be tapping out Morse code as I trace McDouall Stuart's route up the Oodnadatta Track. I won't be communicating much at all. Mobile phone coverage along the Track is a big black hole with splotches of connection in a handful of spots.

The pub's museum display also features sketches of the region by botanist Josef Albert Franz David Hergott, who travelled with McDouall Stuart on his second expedition and spotted a group of springs near where the town of Marree now stands. McDouall Stuart considered that this 'unlimited supply' of water meant there would be no difficulty in taking stock to Chambers Creek at any time of the year. He was so thrilled at this abundance, he named them Hergott Springs, following the habit that many explorers throughout Australia had for ignoring existing Aboriginal names that had been used for thousands of years.

McDouall Stuart's early journals are stacked with creeks, ranges, mountains and springs in Arabana country he named after prominent men, no doubt his benefactors and supporters. There are suggestions, mind you, that some of his name choices, superimposed over Arabana and other First Nations names, were themselves sometimes altered by Chambers or politicians of the day before his journals were officially printed.

The town born near the springs, also called Hergott Springs, was changed to Marree, one of around sixty town names changed in South Australia to eradicate any German influence. Many names were later restored, but not Marree. This mortified one Reverend John Blacket, who wrote in the local newspaper that a self-evident wrong had been done to Hergott and McDouall Stuart. 'That wrong historically must be put right,' he stormed. It never was. Perhaps he wouldn't have got so het up if he'd heard what I was told: that Marree comes from the Arabana word 'mari' meaning place of possums.

Phil suggested I hang around Marree this weekend for the gymkhana, where I'll be sure to meet Dion Khan, a third McDouall Stuart tragic and manager of Stuart Creek Station. The penny dropped. That's the sheep station McDouall Stuart called Chambers Creek throughout his journals, where he'd pegged out a portion he considered a just reward for his exploration work on his first expedition. He'd been forced to return on his second and third expeditions to resurvey the land he'd earmarked, as the South Australian Government shilly-shallied over how much of the land he deserved. I remember pouring over a folder full of letters back in the library in Adelaide, getting all het up on his behalf.

I'm chatting to a punter outside the pub where the Oodnadatta Track presses on into the central deserts of Australia, the dry and dusty landscape that McDouall Stuart rode over, again and again.

'Some say McDouall Stuart went on his expeditions to dry out.' I comment.

'A bit extreme.'

'You're not wrong,' I say.

The camp bed in the cottage is squeaking as I toss around, excited about the prospect of chatting to Dion about Chambers Creek Station / Stuart Creek Station. A beastie I thought was a creepy crawlie has taken flight, battering itself against the walls. Not a peep though from McDouall Stuart.

Wonder if he's annoyed with me for mentioning his drinking.

'Whose side are ye on anyway?'

He's annoyed.

'I see ye've had a couple yersel' tonight.'

I decide no reply is the best response. Another squeak from the camp bed as I turn and face the wall.

NINE

GHAN SPOTTING

I have a front-row seat and a guide well-qualified to show me around for my grand tour of Marree. Just so happens it's the front seat of the local ambulance.

I couldn't think of a better person to while away my time with than June Andrew, then resident nurse at the local clinic. June has been involved in every group, club, team and organisation that bears the Marree label, and I have a feeling only an emergency will stop her from taking me to every notable Marree landmark.

June arrived in Marree in 1982, too late to hear the whistle of the last train leave in 1980 when the track was moved west, but she knows a thing or two about the Ghan railway. We indulge in some Ghan spotting: the street sign 'Railway Terrace', the sturdy railway platform and the old railway engines and carriages; all signs of how much the Ghan once dominated life in this town. June points out buildings with hints of Ghan that only a local would recognise—repurposed sleepers for fencing, dog spikes for paving, railway pegs, rail iron and rail wire. Railway track iron serves as posts under the front balcony of the Marree Hotel—where June is a member of the pub's darts team.

The first mosque in Australia was built in Marree by Muslim cameleers, collectively called Afghans, who played a major part in opening up inland Australia, transporting bales of wool and oil to the railhead and carting everything from furniture to medicine for families living in pastoral stations. The arrival of the motor truck in the 1920s put an end to the strings of camels plodding up the Birdsville, Oodnadatta and Strzelecki tracks. June points out a structure of thatched roof, earthen walls and tree-trunk poles on the main road, a reconstruction of the original mosque.

Onto the Ghan Railway Museum, set up by the local Progress Association—of which June has been treasurer, secretary and president. When the new railway line was built further west, Marree's population dropped to around one hundred and Marree turned to tourism to keep alive—like most other settlements along the old heritage track. We're here to photocopy posters for tomorrow's gymkhana where June's first aid stand will double as the lucky dip stall and the drink-ticket kiosk. I offer to help. I've minimal first-aid knowledge but some experience with drink tickets. I reckon the first-aid stand will be a prime vantage point at the gymkhana to seek out Dion Khan.

We drive past the Country Fire Service, the local school and the kindergarten (June's been involved with them too). Last stop is the local movie theatre, complete with rows of cinema chairs, surround sound and a popcorn machine. June runs film nights once a week.

Neville at the campground was still laughing when he handed

me the keys to a cabin after I'd asked for a patch of grass to pitch my mozzie dome. So far, the swag has been used bush-style only once. My pre-dinner ritual is a constant: cheese, biscuits, olives and a glass of red wine, basically because it involves no preparation, no cooking, and some alcohol. I've realised Bad-Ass Betsie will be sweltering in the midday sun at tomorrow's gymkhana so the remains of the yoghurt, limp lettuce and red capsicum and a lump of very ripe cheese are the making-a-meal-of-it mix for alfresco dinner on the porch. These ingredients have held up well, though others may dispute that. I'm known to take culinary chances. The flies, then the mosquitoes, scupper my plans, so I retreat indoors where I've hauled the car fridge, setting the thermostat to coldest to crank it up for tomorrow.

Next day at the gymkhana, June doesn't require my services after all, but she's sussed out where Dion is for me.

'You can't miss him. He's the big fella wearing a blue shirt and a big hat.' I head to the stand where big hats are abounding. Even the little tackers are swaggering around in ten-gallon creations, watching a foot race with most of the runners minus shoes. On burning sand, that's a tactic to make anyone beat their personal best.

Dion is indeed a big fella. 'What he went through when he travelled through this area... what a man,' says Dion, who tells me the first book he ever read cover to cover was about McDouall Stuart. Dion, Arabana on his mum's side, his dad an Afghan cameleer, started work on Stuart Creek Station as a roustabout when he left school at fourteen, following in the family footsteps back to his great-great-grandfather. He's been the station manager since 2005.

When he's mustering, Dion stops at a hut deep in the property where he reckons McDouall Stuart may have camped. Could Stuart

Creek be the legendary Aboriginal paradise of Wingillpin? I ask.

'Who knows? Maybe is, maybe isn't,' says Dion. 'What I do know is that I wake up in the morning, the sun rises—and I know it's going to be a great day. And when I look around the station—I know I'm in paradise.'

I'm welcome to go there and have a look around, he says. I didn't need a second invitation. I'll be there tomorrow. The station is just up the road. I'll find it.

I cut short my gymkhana outing, head back to Marree and book into the cabin for an extra night, still no luck getting a spot in the shade for Bad-Ass Betsie. Good job she's not high maintenance and seems to be coping. I settle in the roadhouse with a flaky pie and an espresso coffee, plus five-dollars' worth of internet data for a spot of online research for my surprise outing in the morning.

The woman behind the counter tries without luck to get me connected. The two bakers come from behind the scenes, click their tongues and shake flour over their pinnies in sympathy. They can't help either. It's Matt the bush pilot with his slick Gen Y skills who is my hero.

The bakers congratulate me on getting connected. I congratulate them on their fabulous flaky pies.

As I bunk down for a second night, I have an inkling why Lyn and June are still here.

PART 3

OODNADATTA

Expedition Two and Three
April – July 1859
November 1859 - January 1860

TRACK

from Marree
to Oodnadatta

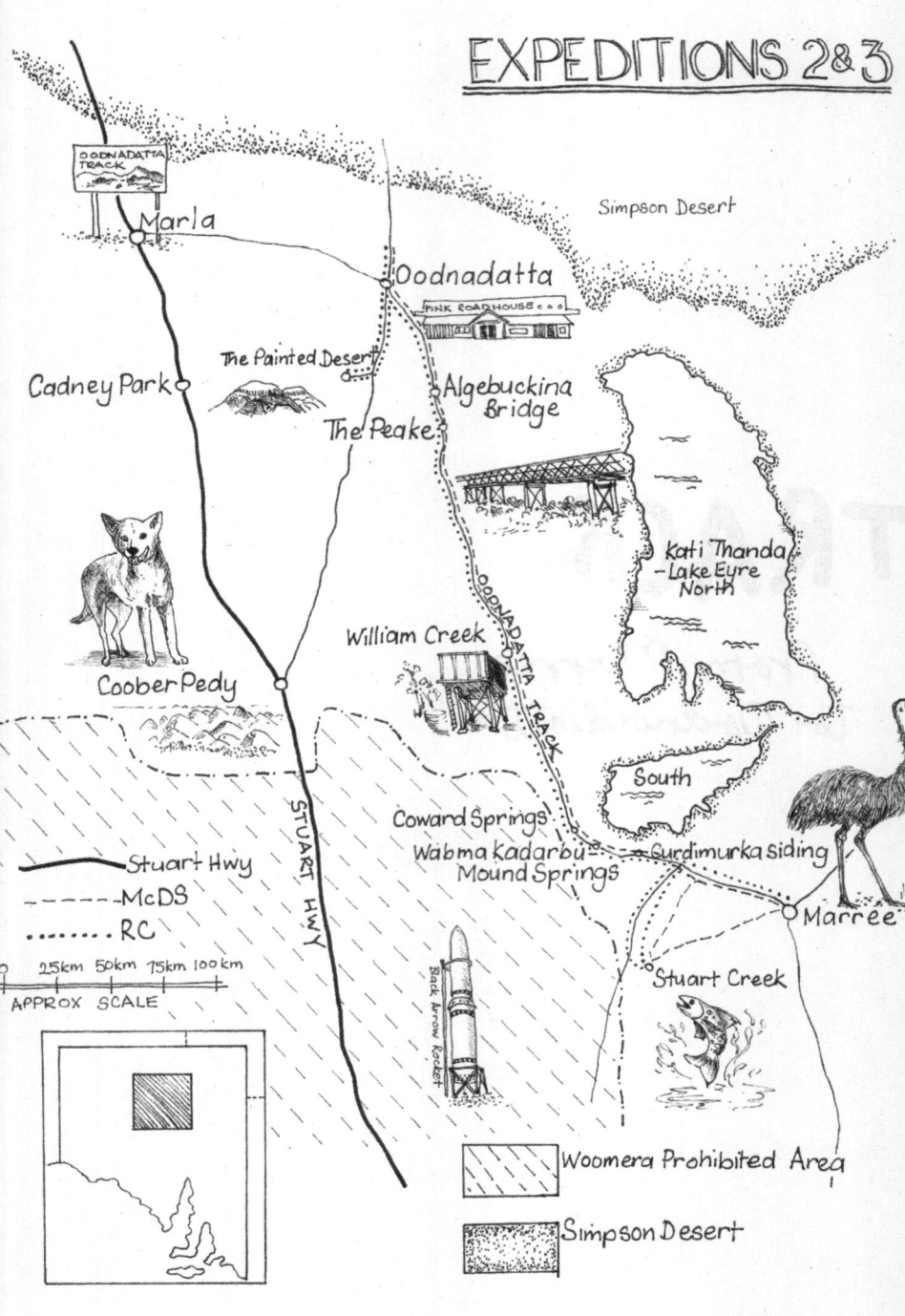

EXPEDITIONS 2&3

OODNADATTA TRACK

Marla

Simpson Desert

Oodnadatta

PINK ROADHOUSE

The Painted Desert

Cadney Park

Algebuckina Bridge

The Peake

Kati Thanda – Lake Eyre North

OODNADATTA TRACK

Coober Pedy

William Creek

South

Coward Springs

Wabma kadarbu Mound Springs

Curdimurka siding

Marree

STUART HWY

Black Arrow Rocket

Stuart Creek

Stuart Hwy
McDS
RC

25km 50km 75km 100km
APPROX SCALE

Woomera Prohibited Area

Simpson Desert

TEN

ENTERING THE ARID LANDS

A no-bullshit sign materialises at the far end of Marree township, like a bouncer appearing outside a rough pub.

CARRY ADEQUATE FUEL, WATER, FOOD, CURRENT ROAD MAPS, TWO SPARE TYRES, TWO JACKS, SHOVEL, FIRST AID KIT, TOW ROPE

Fine soil gusts across the unsealed road, smothering bushes in red dust. This is the start of the Oodnadatta Track and my entrance into South Australia's Arid Lands proper, where temperatures can top 50 degrees Celsius. It's late spring and heating up, the tail end of when it's sane for travellers to venture forth. The next refuelling point is 200 kilometres away... I'm not about to ignore the message.

The list goes on.

AVOID NIGHT DRIVING WHEN WILDLIFE AND LIVESTOCK MAY BE ACTIVE. REMAIN WITH THE VEHICLE IN THE EVENT OF

MECHANICAL BREAKDOWN. ADVISE LOCAL POLICE

BEFORE TRAVELLING OFF DESIGNATED ROADS

Bossy.

There's more.

FLASH FLOODS CAN OCCUR WITHOUT WARNING.

AVOID WET WEATHER DRIVING — ROADS CAN BECOME

DANGEROUS AND IMPASSABLE

The stunted saltbush lining the soft edges of the Track wobbles in the heat haze. Three months ago, mind you, it was a different story: the Track was closed after heavy rains, stranding tourists and marooning communities for weeks.

A second sign tells me the Oodnadatta Track is open for business. I'm on my way.

The road grader's been through, knocking the crests off the corrugations, and the trick is to find the sweet speed to skim over the still-rippled surface to soften the vibrations. Too slow and it's a bumpy ride over those brick-hard humps, loosening bolts, nuts, screws, teeth; too fast and you find yourself enlivened somewhat with an unexpected sideways drift towards a sandbank.

Are 'dull' and 'desolate' the words you conjure up with the words Arid Lands? I was that bunny. But there's nothing boring-beige about this landscape. It's a glorious testament to the brilliance of brown: gritty red-brown sand, strips of brown-black gibber stones gleaming from nature's varnish of manganese and iron, and splashes of almost-black flint stretch left and right.

The brilliance of brown didn't thrill McDouall Stuart on his expeditions. He was on the lookout for signs of a few more springs like the ones Hergott had discovered.

I imagine him mocking me for being so sappy. He had no time to get syrupy about endless flat plains leading to the edge of the world. No. His journals are sprinkled with comments on how many days they'd gone without water, about the piddly dribbles they came across, the great disappointment at finding rivers that had turned into a series of brackish ponds, and water he described as 'saltier than the sea, and of no use'.

Bad-Ass Betsie is a beauty, shooting along the camel-coloured track, edging up to ninety kilometres per hour. Such a capable lass and we work well together, though the rattling in the fridge is disconcerting. 'I promise I'm grateful,' I judder, remembering that McDouall Stuart was pleased if he managed to cover forty miles (about sixty-four kilometres) on a good day.

'Try to dae this on horseback. And nae track, so dinnae complain aboot the corrugations.' He's so predictable.

'I already said I'm not complaining, Mr Stuart.' Why would I complain? I'm in an air-conditioned car, cocooned from flies and the heat, and there's cool water in my bottle. It even has a spray nozzle to mist my face. The only problem is one crabby passenger.

'Cool water,' he grunts. *'The stomach cramps I had wi thon water back when. I could tell ye a few stories. The temperature o' the water wis the least o' ma worries. I wis doubled up in pain mony a night.'*

Time for Bad-Ass Betsie to come to my rescue. She's good at entertaining and some music is needed. No idea how the sound system operates from the steering wheel but I get a CD playing. If McDouall Stuart is still whingeing I can't hear him.

Australia's Dog Fence crosses the Track at the fifty-kilometre mark. It's the longest fence ever constructed anywhere in the world, rolling out over 5,500 kilometres, and they say it's visible from outer space. Yet I missed it. The bone-shaking sensation where the fence line crosses the Track and becomes a metal grid should have been a hint. The wider-than-normal spaces are to stop dingoes padding from the north to harass the sheep in the south.

However, I do notice the strips of tell-tale river red gums and coolabah trees on either side of the Track announcing another creek bed and I brace myself each time as the dips in the road stir things around the car. The big yellow floodway signs are a giveaway too.

I've crossed Welcome Creek, Kidnimpiri Creek, Wangianna Creek, Screech Owl Creek and, my favourite so far, Tinta Dintana Creek. No need for me to ford a raging river or even a babbling brook. The crusty creek beds, dry as a dead dingo's donger (Aussie alliteration at its best), look like they've never experienced a trickle of water in their life.

Flood markers rising to two metres tell another story. The average annual rainfall in the arid zone is 200 millimetres (eight inches)— oh the mockery of statistics. Years can pass without one droplet of rain plopping on the track, and then a few years' worth can fall all at once.

At times such as those, travellers who've ignored that warning sign back at Marree not to camp in creek beds could find themselves gripping the sides of their swags as they're swept downstream on an impromptu waterslide. Even with a more sedate downfall, unsealed roads become slippery. Think wet clay on a potter's wheel. Locals have no time to ponder the false picture of those averaged-out water tables. They're too busy mopping up broken fence posts, twisted wire and drowned animals.

At the 85-kilometre mark, a sign points to a lookout for Kati Thanda Lake Eyre basin, the lowest natural point on the continent. My puny binoculars reveal a glaring-white expanse stretching into the distance, the same massive glistening salt pan that McDouall Stuart saw when he passed through in the mid-1800s, then used in the mid-1900s to attempt land-speed records—notably Donald Campbell breaking the 400-miles-per-hour record on the lake in 1964.

I'd spotted a boat under the eaves of the Lake Eyre Yacht Club back in Marree, which I assumed was a tongue-in-cheek tourist attraction. But no. The club was formed in 2000. Monsoon floodwaters travelling along the river systems of semi-arid inland Queensland a thousand kilometres away have a slow journey south, reaching the lake weeks later, if they haven't evaporated along the way. It's easy to see why the lake fills up completely about once a century, the most recent occurrence being in 1974–76.

But there are decent flows every few years when the lake has water and hundreds of thousands of waterbirds use the islands that form in the middle as breeding grounds. Tourists fly over to marvel at the sight and dedicated sailors from the yacht club float their boats.

This lookout, however, is as close as I'm going to get. There are other options for those with a four-wheel-drive vehicle and remote-area driving and camping experience, but I've studied the brochures from official sources on outback safety. They make it clear you cannot claim you've not been warned. Excitable blog posts from fellow travellers who've come a cropper suggest not to do what they did. I'll come back one day and fly over.

This lake made liars and fools of early explorers, government surveyors and private land speculators. They were accused of believing a mirage or making it all up. McDouall Stuart returned to the lake shores at different points, viewing it from the south and west, and his journal entries describe his attempts to ride and walk on the surface '...the ground became soft, and the clay was so very tenacious and my feet so heavy, that it was with difficulty I could move them, and so I was obliged to return.'

At the 100-kilometre mark, I spy a long skinny building with mean little windows. This is Curdimurka (Kudna-Ngampa), described as the best-preserved siding on the Old Ghan Railway Heritage Trail, the only one still intact. A couple of spindly trees frame a peeling facade only a railway buff could love.

I'm about to drive off when a massive tumbleweed rolls into view and I watch as it skips over the stony ground, bumps alongside the building's veranda and tries to nudge its sizeable girth into the doorway; a perfect symbol of desolation. Then it's off again and I swear I heard Bad-Ass Betsie let out a sigh of irritation as I switch off the engine and step out to follow that bouncing ball. I know my tendency to stop-start, take copious photos and get distracted can be tiresome for a human, but I didn't expect a vehicle to complain. The tumbleweed flips and flops along the railway tracks, heading for the water tower that, once upon a time, fed the steam trains that chugged along this line.

The trackside sign tells me that Port Pirie is 332 miles south and Alice Springs is 477 miles north. The tracks are so shiny, I find myself checking left and right even though I know the last train passed by thirty-odd years ago. A smattering of bickering corellas have congregated around the downpipes that poke out from the disused water tower. It's the only sound within cooee. Not for the first time, I feel I'm the only person on this Earth.

The sun's belting down as I head back to read the Old Ghan

railway sign. The paper is peeling and the writing is flaking, but I can still read bits through the crazed, glazed lines '... a story of outback dreams and heartbreak... in a harsh land'.

A timetable shows the railway line reached Oodnadatta in January 1891, extending to Alice Springs in 1927. In the 1940s, wartime saw traffic go from three to fifty-six trains a week. When diesel locomotives were introduced in 1954, many water points fell into disrepair. A fettler's life was not a comfortable one, and those at Curdimurka battled sand drifts, rail-buckling heat and devastating floods to maintain the railway track. When the entire line was shifted west in 1980, this became one of many abandoned railway buildings.

I dash over to the thin sliver of shade under the building's low-slung veranda and peer through the streaked windowpane. The warm wind whips against my legs as I walk around the corner. I swear I can hear the clink of wine glasses and ghostly music, the faint rustle of evening gowns. For here is the site of the Curdimurka Outback Ball. In the 1980s and 90s, partygoers would flock here, parking their jeeps and landing their light aircraft alongside the railway tracks, before heading for the dance floor laid out under a gaily lit marquee. I'd love to do some boot-scooting at the Curdimurka Ball, but this black-tie fundraiser for the Ghan Railway Preservation Society is no more.

It's not even midday and the sweat is sliding down my back. I jump into the driver's seat with a movement that I'm perfecting to scupper the intentions of those flies trying to follow me. 'See. Worth stopping for, wasn't it, Betsie?'

I spot an emu lurking behind a spindly bush so I slow down. I don't want to play long-legged chicken with her. There she goes, tippy-toeing across my path, ballerina-like in her tattered feathery tutu. She's the first encounter with a beating heart I've met so far on the Track.

The dust in the distance heralds an approaching vehicle. You'd hardly think it possible to see the driver's index finger flick off the steering wheel as we pass each other. But I do. It's the outback equivalent of a handshake and I enjoy the brief connection. The blinding dust swirls around for a few seconds. Then I'm on my own. Once more.

ELEVEN

THE LAND THAT CHARMED

A long skinny pole holds up a rusty oil drum, piercing the unbroken line of the horizon, bright blue above, bold red below. I'm on track to visit Dion. That drum is the post box for Stuart Creek Station, so this fork in the road must lead me to the land that charmed McDouall Stuart.

I drive past another outback icon, a huge water tank, battered and pockmarked, no longer capable of storing water—should it ever again rain out here.

As the dust whips across the cracked claypan, the hot, dry smell seeps through the air vents and I flinch each time a smack of fine red soil hits the windscreen. It's mesmerising flat nothingness all around until a patch of grey-green appears in the distance. Getting closer, it looks like someone has scattered yellow tennis balls through it.

It's still a shock to feel the wind's blowtorch force when I nip out to collect one of those balls. They're fruit attached to a creeping vine and I wonder why this pretty crop that looks so yummy is not harvested, so I pop the evidence in the door pocket to ask someone later.

Dion Khan's home, dwarfed by a cluster of industrial-sized sheds, lies at the end of a long driveway. The usual picture: heavy-duty pieces of farm equipment, their noses poking out, looking primed for action. But there's no sign of life. Even the flies have given up battling the wind.

The third time I knock, Dion nudges the door open, holding it with both hands. His directions to the old stockyards and home-steads down by the creek are blown away by the wind bellowing past him to collide with the footy commentary on the telly in the back room. 'Make sure... road... you... to the right... not...' He juts his chin yonder and points out dips and bumps to lead the way but I don't see the contrasts, the shapes that he must be seeing in the landscape.

Finding the road turns out to be a cinch. It's wide. Really wide. Smooth too. Those tyres lining the sides. How hospitably handy. Why Dion didn't mention this is puzzling. I come to a broad semi-circle forming a neat cul-de-sac, big enough for a small plane to turn around... I'm on the Royal Flying Doctor Service airstrip. Oops.

Dion and his mate are standing on the veranda when I return, putting paid to my plan to dart past the homestead and forget about this side trip. It appears my wheelies are more interesting than the footy. The pair jump down the stairs, pushing their cowboy hats firmly on their heads, and head for their ute out front. Dion nods to me. No words are spoken. I follow in Betsie, thrilled that they're leading me deep into McDouall Stuart country, petrified

I won't be able to keep their ute in sight as they take me along a narrow, twisting, bumpy, unsealed road down to a gully and into an era long gone.

Fences of forked tree branches and wire supporting the crossposts are remnants of stockyards built by early settlers. Chimneys identify two ruins, like the teeth of human remains, one homestead dating back to the nineteenth century, the other from the early twentieth century. Tumbledown walls have disintegrated back into the landscape, the bits left standing threatening to collapse into a pile of rubble. We agree, without any mention of my diversion up the airstrip, that it's best if I don't try to find my way to the hut Dion had mentioned yesterday.

Dion saunters back to his ute. 'You look around here, go for it,' he says over his shoulder. 'If you need us, sing out. Channel 8.' Dion has a footy game to get back to. 'Oh, and watch out for snakes.'

Lyn had given me a wee tutorial back at Wilpoorinna on how to recognise which direction a snake is taking from the shape of its wavy track. It's but a hazy memory. Stomping, I've been told, creates vibrations to encourage any resident snake to slither off and I have the satellite phone in my backpack, confident it won't be needed. 'Coming through,' I call out. 'It's only me.' That's an added touch to signal my approach as I snap and crackle over fallen leaves, twigs and branches in the dry creek bed.

For snake bites, I'd read somewhere that an Aboriginal solution used to be to dig a hole and cover the patient up to their neck for a few days to make sure they kept still, which is similar to today's considered wisdom of wrapping the wound firmly to prevent movement.

Beyond the homestead ruins is a dry, sunken strip, taken over by trees and bushes pushing through the sandy banks. It's nothing like

McDouall Stuart's glowing report of the waterway he'd discovered and named Chambers Creek on his first expedition. That's probably further on. He wrote about long reaches of permanent water and a waterhole 40–50 feet wide. 'The water is excellent, and I could see small fish in it about two inches long,' he jotted down.

On his second visit, to resurvey the land, he ramped up his praise. Those waterholes were not holes, they were long ponds, he recorded. And that creek? 'I am sorry I did not name it a river in my former journal,' he wrote.

'Aye, I wis fair taken wi' the land I huvtae say. I mind gettin Hergott tae sketch the fish ploppin up tae the surface o' the clear water.'

We continue walking along the crackling creek bed.

'Anither thing I mind, wis a shrub that tasted and smelt like cinnamon. We happened tae stir up the sugar in a pannikin wi' a wee twig from the bush and it left quite the flavour o' it in the tea.'

What's on my mind at the moment is that folder of letters back in the Royal Geographical Society library in Adelaide. The label on the front stuck down with aged-brown sticky tape reads: 'Rewards granted by the SA Government to John McDouall Stuart in recognition of his services.' That showed promise that McDouall Stuart's efforts had been acknowledged. Not to be.

The contents of that folder, fat with letters, papers, parliamentary debates and maps, reveal a litany of claims and counterclaims: a catalyst for umbrage from McDouall Stuart's corner and causing unseemly bickering within the fledgling South Australian government. There were concerns that the parcel of land McDouall Stuart wanted was the prime section along the creek. Why should one person get that, the politicians queried. Then doubts were cast on his maps and reports, questioning where he had been and his estimation of the quality of the land. No politician of course left his cosy seat in parliament to check it out. Others huffed and puffed and questioned whether they were obliged to acknowledge his efforts at all.

'You must have been cheesed off at that palaver over your claim on the land.'

'No' just that. They had some glaikit ideas about how those stations should be run. They wanted tae keep leases short, not just for me, but for ither folk tae. Stupid idea. The longer ye can hold a lease, the mair likely ye are tae improve it. And the stocking rates, they were too high, far too high for what that land could cope wi'.'

By this stage, ructions were surfacing over a questionable deal involving McDouall Stuart's financial backers over a mine near Moolooloo. He wouldn't hear a bad word said about Chambers and Finke, so I didn't like to suggest that might explain the lukewarm reception from some members of parliament.

The arguing back in Adelaide dragged on for so long that, in the meantime, bits of the land McDouall Stuart asked for and resurveyed were divvied up and handed to others, forcing him to undertake yet another survey and a subsequent redrawing of boundaries.

Eventually, McDouall Stuart was granted a piece of the land and it became the setting off point for all of his future expeditions. It was listed as Stuart Creek Station, but McDouall Stuart insisted on calling it Chambers Creek in all his journals.

TWELVE

STRING OF SPRINGS

McDouall Stuart was ecstatic with every new spring he came across.
A handful here, a dozen there. 'From whence do they derive their
supply of water?' he queried in his journal.

In this unforgiving land, he knew the dire consequences for horses
and men if they didn't find water. For me, the mound springs were
a novelty, not a necessity for survival but a must-see on my bucket
list of touristy sights.

The title 'mound springs' has stuck though not all springs are
raised. Some are at ground level, others seep from slight depressions
and my camera is full of photos of pink and white swirls across the
landscape. The mounds that rise from the flats, resembling baby
volcanoes, are impressive. Some are lanky, some squat, and many
have sedges and rushes, like wild hair, sprouting out the top.

McDouall Stuart's days spent searching for more and more springs
were long and tough. For me, the boardwalks within the Wabma
Kadarbu Mound Springs Conservation Park made visiting them a

simple saunter. The flies were the only hardship. The park has been cooperatively managed by the South Australian government and the Arabana people since their Native Title rights were formally recognised in 2012, and land management decisions are guided by traditional knowledge and accumulated research.

Thirrka (Blanche Cup) is a classic mound spring, some metres high, while Pirdalinha (The Bubbler) is a glassy pool of clear water at the end of a boardwalk. I arrive as a bubble stirs up the bed of sand and the foam begins to swirl. I have the camera at the ready.

A sign nearby says that in pre-colonial times, Arabana elders whispered a special chant to encourage eruptions, reputed to rise a metre or more into the air. This one's more of a bubble-plop, it's so petite, and I spend an inordinate amount of time waiting for the next eruption, and then the next. Just one more, I'm muttering, like a gambler betting on the next roll of the dice. But all I get are pocket-sized plops and each shot I take has photo-bombing flies creating hazy UFO-like smudges.

According to Arabana legend, Pirdalinha is where the hunter Kakakutanha pulled a serpent from the ground after tracking the strange slithery beast through the region. The death throes of the serpent after being speared are responsible for the writhing in the water. Thirrka is the large ground oven where Kakakutanha cooked his catch and Wabma Kadarbu (Hamilton Hill) nearby is where he tossed the serpent's head over his shoulder. Wabma Kadarbu, about forty metres high, means 'snake's head'. I half-close my eyes and convince myself I can see its profile.

I love the next bit of the story: it gives a whole new meaning to the rootin' tootin' Curdimurka Outback Balls once held back at the railway siding. Kakakutanha's wife, not at all happy at being left to dine on a mere rib bone, put a curse on her greedy husband. Kaka-kutanha's testicles swelled and had reached enormous proportions by the time he staggered in pain into Curdimurka. He beckoned all his people to gather around, whereupon his testicles exploded and hurled into the distance, becoming two little islands in the middle of Lake Eyre South.

Still on the search for springs, I head for Strangways further up the Track, wandering from extinct mound to extinct mound, stumbling upon some more springs that have survived. On and on I go, snooping around for yet another and another: Reedy Spring, Waterfall Spring, Saltbush Spring. I feel I'm channelling McDouall Stuart for sure.

It's been a hot and sweaty day and a hot tub is on offer at nearby Coward Springs (named after a nineteenth-century expeditioner, not referring to someone's personality). An oasis in the driest region of South Australia.

I grab the opportunity for a refreshing soak: even the kiddie's pad-dling pool at Moolooloo in the Flinders Ranges had been drained. I dangle my feet over the side of the rustic mini pool to test the temperature; around 28 degrees Celsius, maybe not that refreshing, but it's wet. That's enough for me.

'You could say this is part of the inland sea all you explorers were searching for,' I comment when McDouall Stuart joins me. 'It was here all the time... under your feet... discovered twenty years after you came by.'

'Well, I'll be blowed.'

'It was above ground tens of thousands of years ago. Now, it's one of the largest underground lakes in the world, below nearly a quarter of Australia from Queensland into South Australia. It's called the Great Artesian Basin, GAB for short.

'I can explain the springs to you too if you like.'

'Aye, they had me flummoxed.'

'I think it goes like this. Rain falling in Queensland percolates down into porous sandstone aquifers and the water moves more or less south-west through the Basin at only one to three metres a year, taking up to two million years to get here.'

'Nae wonder we had stomach cramps drinkin it.'

'When it does arrive, the water hits the rock, finds weak spots, rises to the surface and spurts out... like taps under pressure. At least it used to. Now it's more of a dribble in places.'

'Whit happened tae the ones I wrote aboot in ma journals? I kent fae the start that they were goin tae be important tae get through that land.'

'Elizabeth Springs, the ones you reckoned had "enough water running to drive a flour mill in two or three places"? They're reduced to a series of small seeps.'

'Shame.'

'The Spring of Hope was the one you considered "of the utmost importance." You said it would mean you could come up and down from Adelaide, in any sort of season, and know you'd have a source of water. Well. There's contention about whether that's still visible and flowing.'

'Cannae believe it's changed sae much in such a short time. Whit happened?'

'Well, the GAB was treated like an underground water tank, expected to constantly refill itself. Not clever.' More than 10,000 bores have been drilled into the Basin for irrigation, mining, town water and pastoral stations and it's estimated that natural flows have dropped by a third.

'As well as the number of bores, many weren't looked after and a lot of uncontrolled water was wasted. It flowed over the land, soaked into the ground, and got lost to evaporation. And so the water pressure dropped. Springs died, others were reduced to spurts and slow seepages.'

The pool we're dangling our feet in started as a government bore for trains and the railway settlement that sprung up. As with many bores, the water wasn't properly capped and also the salty water corroded the bore casing, so by the 1920s millions of gallons of water flowed over the gibber plains, creating a natural pond. Railway passengers and locals loved it.

The bore was rehabilitated in the 1990s and the flowing water is now channelled into this pool and then into a wetland providing water and food, shelter and breeding areas for all kinds of wildlife.

THIRTEEN

AN ICY COLD BEER

The William Creek Hotel slides into view: a long, skinny building held together with more than its fair share of corrugated-iron patches. That offer of an icy cold beer is a grand idea.

The shouty advertising along the length of the veranda isn't to fight off competition. This is the only pub in town, the only pub for hundreds of kilometres in any direction. It's taken me three hours to drive the seventy-four kilometres up this deeply-rutted section of the Track this morning, Oodnadatta is 200 kilometres further on and the pubs in Coober Pedy in the west are at the end of a 165-kilometre side road—usually passable, not always.

McDouall Stuart was rather upbeat when he came through here on his way back south on his third expedition. He knew he was onto something with the string of springs he'd plotted on his map. His journal entry reads: 'By the discovery of springs on this trip, the road can now be travelled... and not a night without water for the horses.'

Twenty years later, this spot had become a support station for the camel drivers bringing supplies for the construction of the Overland Telegraph Line. The Ghan railway line arrived about ten years after that and the hotel, now state-registered, made its appearance, initially as a boarding house. The current dining room is made from railway sleepers left behind when the track was moved west.

William Creek, halfway between Adelaide and Alice Springs, claims to be the tiniest town in South Australia, mustering ten residents at the last census. It is also smack in the middle of Anna Creek Station, Australia's largest working cattle station, half the size of Tasmania.

The street is empty except for a truck with a blue water tank bolted to the tray top. Someone splashed spare white paint on the cab door and over the tank to declare its role as the tiny town's fire brigade. The lonesome parking meter alongside has an expired flag, rusted on, there for a laugh.

All the action is in the front bar where I'd swear the cast of an Outback soap opera has gathered. A fit-looking cowboy taps his broad-brimmed hat as he chats to his sheila, squeezed into skin-tight jeans and a figure-hugging, tasselled shirt. They look ready for a rodeo. Or a photo shoot. Next to them is a pair of dusty, sweaty tourists with a puncture and lots of questions for the man behind the bar. They're going nowhere in a hurry: I overhear spare parts will have to be brought in. They might as well have a latte, book a room and relax. The only space at the bar is next to two old-timers in the corner, settled in for the day, who are forecasting a windy night to anyone who'll listen. An oppressive nor-westerly they tell me.

I've walked into the town's internet cafe, supermarket, tourist information booth, post office, pay station for petrol, starting point for all problems car-mechanical and booking hub for accommodation, flights and tours. The décor makes it feel even busier, with business cards, scraps of poetry and philosophical musings scrawled on notepaper covering the ceiling and walls. Hats, shirts, knickers and G-strings are in your face. I spot an Irish hurling stick in the corner.

The shiny-haired gal behind the bar reels off prices for the various accommodation options and hands me a key to check out a cabin before I hand over my cash. She tells me she's from Tennessee, came to this country as a backpacker, met and married an Aussie bush pilot, and they're spending their honeymoon travelling around the outback. How adventurous. Then she turns away, reeling off the accommodation prices to a new punter.

Over at the camping ground, I ponder whether I want to spend the night in this metal box. A swinging light bulb scans over the narrow bed that takes up half the floor space. The one power point is high on the wall but I reckon if I perch the car fridge on the stool in the corner, its cord should reach. I have a running battle with that fridge. Alternatively, I could try to hammer tent pegs into concrete soil but the wind is swirling around and shifting anything not tied down. The decision is easy. Metal box it is.

Cooking dinner is not so easy. No matter where I set up the stove outdoors, the hot wind pursues, and I don't have enough hands to pin everything down, so I go to bed, convincing myself I'm not that hungry. That fridge at my earhole and I are having a rumbling competition.

I sleep through the windstorm which the Tennessee gal tells me the next day had a couple chasing their tent and lashing it down with extra ropes in the middle of the night. Am I glad I chose the metal box.

A rumour of approaching stormy weather is rippling along the bar. Everyone has an opinion. 'Aye—always rains after a nor'westerly,' says one confident predictor. I'm hoping he's wrong. Or at least that the rain holds off. Selfish thought I know, but there are lots of blue squiggly lines on the map and I don't want them to materialise into running water across the Track. What's a few days more for folk who've been praying for water for years.

The same two likely lads from yesterday are back at their spot at the end of the bar. Brian is a hippie/hip-hop combo, with flowing locks, now snow-white, a remnant from the 60s and an earring that has stretched a hole in his earlobe enough to poke through a sizeable stick. He appears to be the all-round rescuer and helper in these parts. His mate Norm is dressed in the style of an R.M. Williams advert; blue shirt, moleskins, sturdy boots and wide-brimmed hat. A perfect pair for advice about my developing desire to 'discover' more springs in an area that would test my four-wheel-drive skills and Bad-Ass Betsie's stamina. The ruins of Old Peake Telegraph Station, halfway between here and Oodnadatta, are at the end of a rough track referred to as PAR 12. PAR stands for Public Access Route. I'm finding 'access' is a word with many possible interpretations. The cautionary note in the map book is that an hour is needed to cover this thirteen-kilometre stretch. Should I attempt it?

'Have you got a UHF radio?' Brian asks.

'Yep. And a satellite phone.' I make a mental note to check later that it's charged. 'And an EPIRB,' I add, remembering the contraption in the glove box. Untouched. EPIRB stands for Emergency Position Indicating Radio Beacon, a device mainly for sailors in dire straits on the ocean blue, and only to be used as a last resort. I've no intention of touching it, but I reckon I spot a nod of approval.

"Bout three weeks ago, two women and five kids had a puncture four kilometres up the road,' says Norm. 'That was when they realised they'd left their jack back in the garage in Melbourne.' I tut disapprovingly.

'They set off with a bottle of water and a bar of chocolate.' Am so hoping he saw me buying a ten-litre cask of water.

The mere mention of optimal tyre pressures has the entire pub chipping in. You could write a thesis on outback opinions about tyre pressures. 'Drop the pressure, no brainer.' 'Too low, and you'll burst the side of the tyre.' 'Better than getting bogged.' 'No fun if the tyre rolls off the rim.' 'It's rocky in parts, pressure should be up a bit.'

'Let's have a look,' says Brian, and we turn our backs on the rubbing of chins to inspect Bad-Ass Betsie. Of course, she got the tick of approval. I'm not doing so badly myself.

Four kilometres up the Track, I spot a car at the side of the road. I blame that tick of approval from Brian for what happened next.

Two guys on their hunkers are inspecting the space that once housed the rear left tyre. 'I've got a fancy pump-up jack you can use,' I call out.

'Got one, love,' one says as he rises to his full, burly height and nods to the rear of the car. What was I thinking? It's full of gizmos. They'd know all about dual pistons, rollers, two-piece handles and swivel castors. I walk over to see the buckled wheel rim and a sliver of rubber that was once a tyre and conjure up images of a shoot-out. Or a land mine. 'I've never seen a puncture doing that before.' You'd think I'd know when to shut up.

'I have.' A man of few words.

Having enough horseshoes was an issue that McDouall Stuart

never seemed to get quite right. I had no intention of making that mistake with tyres.

After Rob in the car hire yard back in Adelaide had introduced me to my outback partner Bad-Ass Betsie, I'd pestered him with queries mechanical right up to the departure date. He had a resigned look about him on pick-up day when I made a request for a rundown on what to do if I got a puncture. I should stand at the side of the road, stick out my leg and wink, he said. For fuck's sake. It was a serious question, I told him, and I insisted on a trial run to disengage the spare tyre lodged under the chassis. I soon regretted that demand. Getting the floppy rod through the hole above the number plate wasn't easy, and poking and prodding it into the darkness beyond was nigh impossible.

'You'll be fine when no one's watching... when you're out there... In the middle of nowhere... on your own...' Rob was trying to be encouraging, but he insisted on wedging a second spare wheel behind my passenger seat for easy access. He then gave me an underbelly inspection of the jack points, which I failed to spot, but I nodded anyway. He was the one who put that EPIRB in the glove box.

Why were ye thinkin aboot horseshoes back there? McDouall Stuart's ability to read my mind is unsettling.

'Well. You didn't take any spare ones at all as far as I can see on the first expedition. On your second, you wrote, and I quote, "We are now come to our last set of shoes for the horses". You then wrote "Having experienced the misery of being without them in my previous journey, I am, though with great reluctance, forced to turn back".'

'All right, all right.'

FOURTEEN

CHAMPING AT THE BIT

I spot a handwritten sign to Freeling Springs and the Peake heritage site, scoured by sand, scorched by the sun, and the scribbles, pretty in pink, all but obliterated. Then I spot the government sign. It's imposing, befitting a destination considered of state and national cultural importance. Can't say the same about the road. It looks rough. Betsie, are you ready for this?

McDouall Stuart was more than ready to survey this land. After this commitment, he could get onto the serious business of ridgy-didge exploring. Three times he'd set out from Adelaide: each time he was constrained by instructions to survey land for pastoral interests and scout for gold; each time he craved the freedom to head into the centre and beyond; each time he pushed the boundaries. He must have been champing at the bit.

My map says I can expect frequent washaways, jump-ups and severe corrugations on this Public Access Route, which is not maintained. When Bad-Ass Betsie lurches and the ground rises, I can't say I wasn't warned. I have to remind myself that Brian back at William Creek was confident we both were capable.

The threat of slamming into the soft, sandy bank gets the old heart pumping. Clutch, change gear, release, press, release. Brake. Now this rubbly creek bed is threatening to topple us over. I've travelled ten kilometres so there's no point in turning back, even if I could. Now I'm dipping and diving like I'm a crazy off-road loony called Gibbo or Dicko. I never did watch the action-packed DVDs that came with the four-wheel-drive magazines. I was never going to be this daft.

I skim an upside-down tree, roots flailing cartoonishly in the air, its feathery-grey twigs brushing the ground like a wild witch's hair. Then another, and another. I'm in an upside-down world. An impish willy-willy skitters across the road in front of me, I can see why these mini tornadoes, dust spiralling into the air, are spirit forms in Aboriginal Dreaming. I pass a pile of rubble (the title of ruin is no longer appropriate) and round a corner.

Before me is a carpark, hard-packed earth, orderly, and in very good nick. This is such a contrast that, for a second, I wonder if I've imagined that scary drive I've just completed. A brochure all about Peake Repeater Station and Freeling Springs, housed in a sturdy piece of modern infrastructure, tells me the site has been upgraded, but not the track. You don't say.

Bollards stand erect to corral vehicles into designated areas but there's only one other vehicle here so I can take my pick for parking. I nudge Bad-Ass Betsie as close as possible to one of the few accessible upright trees in sight, brushing under a low-flung branch to secure a smidgen of shade.

Friends of Mound Springs, established in 2006, installed the bollards and the walking tracks to raise awareness of the site's importance and they've been joined more recently by the Arabana Rangers, who work with them on maintenance and conservation work.

Under the influence of all this order, I obey the brochure's instructions to wear strong shoes and I lace up my walking boots. Minimise bare skin, it reads, so shorts make way for trousers and a long-sleeved shirt goes over the T-shirt. I slap on a hat and am slopping sunscreen on the remaining patches of bare skin when a couple walks into sight. It's their car parked in the blazing sun.

'You made it then,' says Andy, who I recognise as the confident weather predictor with the Scottish brogue from the William Creek pub. He badgers me like a protective big brother but I insist there's no need to hang around to escort me back out. But I don't mean it. I so want to follow them out of here, right this minute. But I've come all this way. And spent all this time slipping, slopping, slapping.

Andy and Elaine extract a promise from me to meet them at Algebuckina Bridge on the Neales River later in the day. 'Give us a call when you're leaving,' Andy calls out as I walk off. 'Just so's we can put the kettle on, eh?' A young couple wearing shorts and sandals arrive and follow me along the path, making me look rather overdressed.

I round a corner and spy a pointy cairn of glistening stones at the top of a hill and clamber up, glad after all for my sturdy shoes to deal with the loose stones tumbling down. Did McDouall Stuart build that cairn? Doesn't say. Another mound is identified as a pile of fossilised nineteenth-century dung, produced by a herd of goats. From on high, I watch that scantily-dressed pair skip around the paths and leave without finding out that little nugget of information.

I see no sign of the Freeling Springs, which McDouall Stuart described on his second expedition as the largest springs he had yet seen, recording: 'The flow of water from them is immense coming in numerous streams and the country around is beautiful.' Wonder where they are. He'd been excited too at the sight of quartz, so plentiful it was like a covering of snow, which he knew was a sure sign of gold. His companions said they'd never seen land that so resembled the Victorian goldfields.

However, on his third expedition, returning to this area, McDouall Stuart was less effusive about the potential for prospecting, 'No gold, no gold' was the refrain on this occasion. McDouall Stuart was in a bit of a mood, complaining about the lack of proper tools and not enough men to dig. 'Our tools are getting worn out,' he lamented.

McDouall Stuart recorded finding evidence of Aboriginal people at Peake and he notes that he came across a group who 'made signs for us to be off'. He then writes in his journal: 'I found that we had camped close to a large quantity of acacia seed that they had been preparing when we arrived, but had no time to carry it away before we were on them.'

I'm in no hurry for the drive back to the Oodnadatta Track, so I saunter around the marked walks. I am dressed for it, after all. One building, I read, started as the homestead built a few months after McDouall Stuart's survey work. A few years later, it became the Peake repeater station for the Overland Telegraph Line, built in the 1870s.

Other signs of the settlement include rubble where a police station once stood and the ruins of an eating house called the Hammer and Gad, the universal symbol of mining (a gad is a type of chisel). There's a mine and a smelter. Not for the gold that McDouall Stuart had been looking for, but for copper, which was mined until the early 1900s, not very successfully.

With its wealth of history, pre- and post-colonial, Peake is now placed on the South Australian Heritage Register. One Dreaming story relating to Peake, collected by the late Luise Hercus, a noted linguist from the Australian National University, is about the two springs closest to the Peake called Yardiya Parnda and Yardiya Kupa, meaning big and little spindle. The two ancestral snakes Yurkunangku, the red-bellied black snake, and Kurkari, the green snake, camped at Yardiya and spent a lot of time sitting there making hair string with a spindle. To stop the wind from blowing away the hair they were using, they built a wind-break of rocks, which is seen today.

As I leave, I flick up the UHF radio switch, ready to practise my roger over and outs and tell Andy and Elaine I've been dawdling. I've seen my share of road movies and know the drill. Not a whistle or a hiss do I hear. Flick down. Nope. I press every button and knob I can see on the dashboard and under the dashboard. Not a crackle.

'On my way,' I shout into the speaker.

FIFTEEN

STURDY STEEL GIRDERS

'That was quite an entrance.' Elaine points to Bad-Ass Betsie who is gaily spinning her flashy yellow topknot. The three of us watch as with each revolution she spotlights the UHF aerial—or rather, the stumpy end of it. That explains the lack of a hiss and crackle.

We blame the bumpy track for wiggling off the top of the aerial, but I suspect I'd got up a little too close and personal with that lonesome tree back in the Peake carpark. The flashing light was my doing too, activated with all my switch-flicking around the dashboard.

McDouall Stuart found plenty of water in the Neales River when he came through here and, taking his cue from the channels spread over the dry floodplain for a mile and a half, he calculated the flow could be impressive. 'The drift stuff was upwards of 13 feet high in the gum trees,' he noted in his journal.

I get my clue about the river's potential from the crumpled carcass of Fred's FB Holden. It's not the first abandoned wreck of a car I've seen this trip, but this one slumped near the Algebuckina

Bridge has its own story on one of those pink and white signs I've been spotting, a tale of how flash floods can ferociously change the landscape.

The sign tells me that Fred's creative solution to get to Oodnadatta in the big flood of 1976 was to use the railway bridge, twelve metres above ground and just clear of the rising floodwaters. He was inching his car forward, placing wooden beams along the tracks, section by section, when he heard a sound over the thunder of the raging flood. He wasn't to know a work gang was repairing the tracks and was careering towards him on a rail trolley. Fred and his vehicle were catapulted over the bridge and into the swollen river. The mangled car, now rusty brown-black, is like a goggle-eyed monster, an empty socket where the headlamp once was. Fred survived to tell the tale.

Rusty remnants, tracks and bridges outlining the path of the Old Ghan railway are all along the Oodnadatta Track and I've photographed most of them. I'm excelling myself, taking photos of the Algebuckina Bridge from on high, from below, from near and from afar. It is an amazing piece of infrastructure in such a remote location. It's fitting that I share my hipflask of Drambuie with my new friends as the sun sets behind the sturdy steel girders, which were brought out from Scotland. Completed in 1891, this bridge held the record for more than a century as the longest bridge built in South Australia and is State Heritage listed.

'Feeling right at home, you two?' Elaine humours us as we launch into our version of 'roaming in the gloaming' even though neither of us comes from the bonny banks of Clyde. Nor can we sing.

'Slàinte mhath' Andy and I say in unison as we raise our plastic tumblers to salute the sunset. Pronounced slan-je-var, it's the one

Gaelic phrase I know.

'The railway follows McDouall Stuart's tracks,' I point out.

'Aye, and I believe he comes from the same part of Scotland as you,' says Andy. Appears I may have made this little speech already. Perhaps more than once.

I have a ringside view of the impressive Algebuckina Bridge from my camping spot close to the river's edge. Maybe not the best choice. Not such a good Scottish connection are the sand flies—the Aussie version of the Scottish midge. My arms are coming up in lumps.

Confirmed McDouall Stuart tragic though I am, I'm sticking to my plan to temporarily abandon McDouall Stuart's authentic route. Some weeks ago, I read that the bikes, cars, buggies and quads competing in the annual Finke Desert Race have made a dog's dinner of the track further north. Best not to drive alone in one of the most remote places in the world and joining a convoy is simply not on. It would be too embarrassing, considering the slow speed I drive at. Too bad I won't see that section of the railway line but, hey-ho, how many photos of railway features does one need?

I'll join up with the railway line again in Alice Springs.

SIXTEEN

LUSTING FOR A SHOWER

Oodnadatta shares its name with a crater on Mars. They both have a rusty, red complexion but this is the beginning and end of the Martian connection. Planet Mars is easily seen from Earth, while getting close enough to spot this tiny township requires some planning. There's no public transport and road access isn't a year-round guarantee.

Mars is bitterly cold with an average annual temperature of minus 60 degrees, while Oodnadatta holds the record for the highest temperature in Australia at 50.7 degrees (123.3 degrees Fahrenheit), reached on 2 January 1960. (Like all records this is disputed but let's not let a sliver of a percentage spoil my story and we'd all agree that's bloody hot.)

When the mercury soars in Oodnadatta, media stories abound of kangaroos lying on their backs with their legs in the air, birds falling out of the sky and soft drink cans exploding in car boots. Oodnadatta can get so hot that the petrol evaporates between fuel pump and tank, so no one with an empty fuel tank is going anywhere any time soon. When I arrive, a breeze is keeping the flies at a bearable buzz and the temperature at a reasonable 35 degrees, letting me fill up. Which is just as well, as it's another 200 kilometres to the next petrol stop, at the end of this Track where it joins up with the Stuart Highway.

It's time to give me and Bad-Ass Betsie that bitumen break offered by the Stuart Highway. She's done well so far, without so much as a puncture or a broken windscreen, not counting that stumpy aerial. Even though the Oodnadatta Track is considered the mildest of outback roads, there are rusty cars, rolled and abandoned by the side of the Track, to show what can happen to the unwary. More to the point, my cracked heels need attention and not even my commando flynet can hide that washing my hair in bore water has changed it to straw. I'm lusting for a decent shower.

When he reached the Oodnadatta region, McDouall Stuart was the first white man the local First Nations people had come across. Only a few years later, radical changes began to take place. Pastoral stations were established, the Overland Telegraph Line was built and the Ghan railway track came through. The ancient trade route through this country was being transformed into a colonial transport and communication corridor.

Some say the name Oodnadatta is an adaptation of the Arrernte word 'utnadata' meaning 'blossom of the mulga tree'. If there were mulga trees in the past, there's little evidence of them now. Another interpretation, not so romantic or poetic, suggests the meaning is from the Arabana word 'kunda thata' meaning 'much faeces' and explains the reddish-brown colour of the soil. Arabana elder Syd Strangways says 'kunda thata' refers to a Dreaming story about an old Arabana man, travelling from some way south up to the Macumba river area, who could not control his bowel movements.

Oodnadatta's role in the Ghan's story began when the line extended into town in 1891. The township flourished as the railway terminus for nearly forty years and Oodnadatta became a major crossroad, a stopping-off point for drovers, Afghan cameleers, miners from all over the country in search of gold, pastoralists coming into town and Chinese market gardeners. And so, the mix of cultures in the population expanded.

Drovers came through with mobs of cattle from pastoral stations to put onto the trains and the cameleers met the trains to shift mail, freight and travellers as far away as Alice Springs—a six-day camel ride north. By 1893 there were 400 camels based at Oodnadatta.

Oodnadatta today is home to around 200 permanent residents plus travellers passing through. They all appear to be with me in the Pink Roadhouse, which is listed as a local 'must see'. That's a superfluous promotion as you can't miss it. It's pink for sure, and pink is not the generally used paint for buildings in the outback.

Neville and Adriana Jacobs had been running the Pink Roadhouse for a couple of years when I pass through. Neville tells me the pink signs I've been passing were the idea of Adam and Lynnie Plate, the first proprietors of the Pink Roadhouse. Adam was also the one who named and got the Oodnadatta Track on the tourist map, helping to make it a popular outback route from the Flinders Ranges.

The roadhouse serves also as the local post office, roadside repair shop and service station, grocery store, souvenir shop and outlet for local art, and provides a mail run on Tuesdays and Thursdays to pastoral stations a hundred kilometres further north.

'We've painted it even more pink,' says Neville. I nod to his pink shirt. 'This isn't pink,' he smiles. 'I like to call it tomato red.'

'I'll be staying in the outback,' says Neville, who originally comes from Adelaide. 'I don't miss anything. You can get everything here. It just takes a bit longer. You miss out on nothing.

'You can't get lonely. Every day new people are coming through here. In the city not many people know their next-door neighbours. Same old Groundhog Day for them. Up in the morning, then into the traffic. Here, there's none of that.'

I wouldn't have picked me for a trainspotter or whatever the name is for someone who considers a clump of rusty rail nails to be as photogenic as a garland of roses. I take photos from all angles. So too, stumps of railway sleepers in various stages of decay. Algebuckina Bridge was a major photo fest and I'm still at it on the outskirts of Oodnadatta, down on my hunkers snapping away.

Because I haven't yet had enough of trains, I take up Neville's suggestion to visit the Ghan Museum in the old stationmaster's house.

Grainy black and white photos on the museum walls show men in rolled-up shirt sleeves wielding hammers and pick axes. It's the 1920s and they are working on the 450-kilometre-long section leading north from Oodnadatta and ending at the township of Stuart, now called Alice Springs. Those photos of workers on the Ghan railway line stretching into the distance are full of grit and guts.

The postie, the pilot, the hamburger maker. How many people in the Pink Roadhouse do I need to ask to shore up my decision not to follow the Old Ghan railway heading due north? I'd been warned back in Adelaide about the spikes, those big, heavy, sharp nails that once pinned down the railway line beyond Finke, now scattered and hidden in the sandy soil, ready to puncture my tyres. The Finke Desert Race has a reputation for being one of the most difficult off-road courses in one of the most remote places in the world.

'If you take plenty of water and food; and you know how to change a wheel,' is the response to my question from Pete the postie, as he swings bags of letters and bundles of newspapers out the back of his mail van. I was sure he'd say no.

I rope in the Royal Flying Doctor Service pilot to help me with the internet. I ask him too, expecting a no. 'Reckon so,' says he, looking skyward. 'But I'll get someone who knows about the roads.' I guess road conditions don't concern him much.

He brings back the guy in the kitchen. 'Why not?' says he after checking out Bad-Ass Betsie. Ah. The confidence of the youth of today.

Instead of saying no, they're saying 'go'. The further I travel from Adelaide, the more I meet people who reckon I'm capable. Is this what I want to hear? I know how to change a wheel. Whether I have the brawn to do it is another question: but it's all technique, right?

Pete the postie and I have a coffee while passengers spill out the side of his van for a wander around the streets. He's added this tourist service to his twice-weekly mail deliveries. On my map, Pete shows me his twelve-hour round trip from Coober Pedy, taking in William Creek, a few cattle stations and Oodnadatta.

'Let's be honest. The region here, it doesn't suit everybody. But, to me, it means complete freedom.' Pete, originally from Melbourne, moved up north more than forty years ago. 'This place has a beauty that doesn't hit you in the face. You have to go and look for it.'

The contradictions are what appealed to me on the way up the Oodnadatta Track; the land looks both tough and fragile at the same time. The soil is challenging, with the clay that swells and waterlogs in the wet and shrinks and dries in the dry, but the native plants understand that and have adapted.

Pete is another McDouall Stuart fan and gives his passengers what he calls his Reader's Digest version of the man. 'I tell them that Stuart recognised that the Aboriginal people lived out here, and the land belonged to them,' Pete says. 'He wrote in his journals that they were the healthiest-looking Indigenous people he had seen: fine, strapping men.

'When they threatened him, rattling their spears, screaming and shouting, he'd say to his men something along the lines of "let them know we have more firepower than them, shoot the tree or the ground, but don't shoot the poor wretched men. After all, it is their land".'

I spot Neville and ask if he has a UHF aerial in his workshop to replace Bad-Ass Betsie's stump. If he has, I just might consider following McDouall Stuart's tracks north. He doesn't.

Peter herds his passengers back into the van and I drive off to meet up with Andy and Elaine at Arckaringa Homestead to spend sunset and sunrise taking photos of the Painted Desert. In the morning, I'll follow them to the junction with the Stuart Highway for a relaxing drive up to Alice Springs.

That night at Arckaringa, before snuggling into my swag, I spend time going through photos I've taken along the Oodnadatta Track.

Twice, McDouall Stuart reached close to what's now Oodnadatta, further north than any other explorer, but he was yet to reach the northern boundary of the province of South Australia. On his second expedition he wrote that, if he found water, he'd try to reach the boundary. He didn't find water and turned back. On his third expedition, he was equally unsuccessful.

As McDouall Stuart travelled on those two expeditions, finding spring after spring, the land back south around Chambers Creek and what's now Marree was already becoming well-trodden as colonial pastoralists moved further and further north. The environmental damage was already happening. Governor MacDonnell, who decided to look at the country that McDouall Stuart was talking about, was mortified to find Hergott Springs back at Marree depleted and fouled by stock and horses.

One photo of a sign at Pirdalinha (The Bubbler) gets me thinking. 'Many things changed once the settlers arrived,' the sign in the photo reads. 'The area was trampled and rubbished. A tree next to the spring representing an ancestor in the snake creation story was cut down and used for firewood. The water pressure has altered, and the roars and twisted bubbles have disappeared. People lost control over their country, languages were lost and they could no longer live on their lands. Care for the country became very difficult.'

I wonder how far we haven't progressed in the last 200 years.

Then I read another section on that sign in the photo: 'Lives have changed, country has changed and connections have changed. Aboriginal people still live in this area, and care for it on many levels. Some are rangers, some pastoralists, some teachers and some run cultural tours for visitors. For Aboriginal and non-Aboriginal people, sharing ways of living in and caring for the country continues to evolve.'

'I'm looking at your encounters with the First Nations folk,' I say to McDouall Stuart. 'I can see they were curious at first about you lot and your strange animals. They'd examine your tracks for sure.

But then they became increasingly pissed off with more and more folk coming up onto their land. Didn't you notice?'

'Aye. Hindsight is a marvellous thing, hen. It wis clear one time on the second expedition that they'd seen white men afore. They knew the name for horses. They wur amused with oor equipment and they seemed inoffensive. Unoffended. Followin us up and doon the creek.

'But yer right. On the way back doon on that occasion, we met a few natives. An auld man came up to us and seemed frightened and trembled a guid deal. Didnae stop him from checking oot ma pipe, mind you. And he kept it tae.

'I thocht they wur takin us to some springs, and wis disappointed when they showed us some rainwater in a deep hole. I wrote in ma journal that they wur quite surprised to see oor horses drink it a'. Horses cannae half drink water. Lookin back, I can see the rainwater wis far better than spring water. He wis being generous. Nae wonder he wis scunnered at the horses drainin it. I can see noo why they widnae go ony further wi' us, and didn't show us ony mair water.'

'On the third expedition, they widnae come near us at a'. As soon as they caught sight o' us they'd bolt. I would used tae try and get them to come doon nearer, but they widnae. In fact, they'd make signs for us to be on our way.

'I get the feeling that the third expedition didn't go as well as the second among your companions either.'

'You could say that. Thank God I had Kekwick with me by that stage. He proved to be top-notch. Mair than I can say for Smith who ended up desertin. He wis a lazy, insolent, good-for-nothing man.'

'Yes, you didn't hold back with your criticisms.'

'I sent Kekwick to Moolooloo to get mair men and horses and waited at Chambers Creek, preparing to get out there again.

'I wisnae going tae waste time goin back tae Adelaide. I wis keen to *get goin again. I had a border to cross, I had the centre to see, and I had the whole continent to cross.'*

Much later, during my research, I catch up with Aaron Stuart, a traditional owner and respected elder, to chat about caring for country and Native Title. Aaron is also a director of the Arabana Aboriginal Corporation which, in 2012, celebrated the granting of Native Title to over 69,000 square kilometres of their land. It runs along the Oodnadatta Track, including Kati Thanda Lake Eyre and the Wabma Kadarbu Mound Springs Conservation Park, and stretching north-east of the township of Oodnadatta.

'Subsidiarity is the word I use for the Native Title process,' says Aaron. 'Subsidiarity is an acceptance that Native Title is a whitefella process we have to work with for the common good for all. But we have all had to make sacrifices and compromises.

'That is how government works. What government wants is security: security for industry, mining, tourism. In other words, development. They don't mind if there are a few arguments along the way.

'But it does affect our stories and songlines,' says Aaron, who grew up in a tin shack at Marree, spending a lot of time with his grandparents, learning about the bush and learning the stories.

'Back in those days, Aboriginal people hardly had any voice, because we were ignored, and even the linguists and anthropologists didn't always listen properly and got things wrong,' he says.

'The Native Title process cuts across the different law systems for different groups. And Native Title has created overlaps on the land.

'There have been tensions and conflict. I cry for the people who have passed away in the process, people who have given up their land, their identity, accepted changed boundaries.'

While a determination of the 2012 Arabana claim was reached by consent, it still took fourteen years, and another small triangle of land near Marree took another eight years.

'We must look to the future,' Aaron says. 'It's not just about royalties, compensation for development on our land.

'There is fractured unity, that's what I call it, but my job as an elder today is to remind people about the land. The Arabana Rangers group, for example, and its relationship with the volunteer group Friends of Mound Springs, government bodies like the EPA, and all who have the same desire to care for the land. My message is: 'it's about sharing culture, sharing truths and caring for our land.'

PART 4

INTO THE

Expedition Four
March – September 1860

CENTRE

from Oodnadatta
to Attack Creek

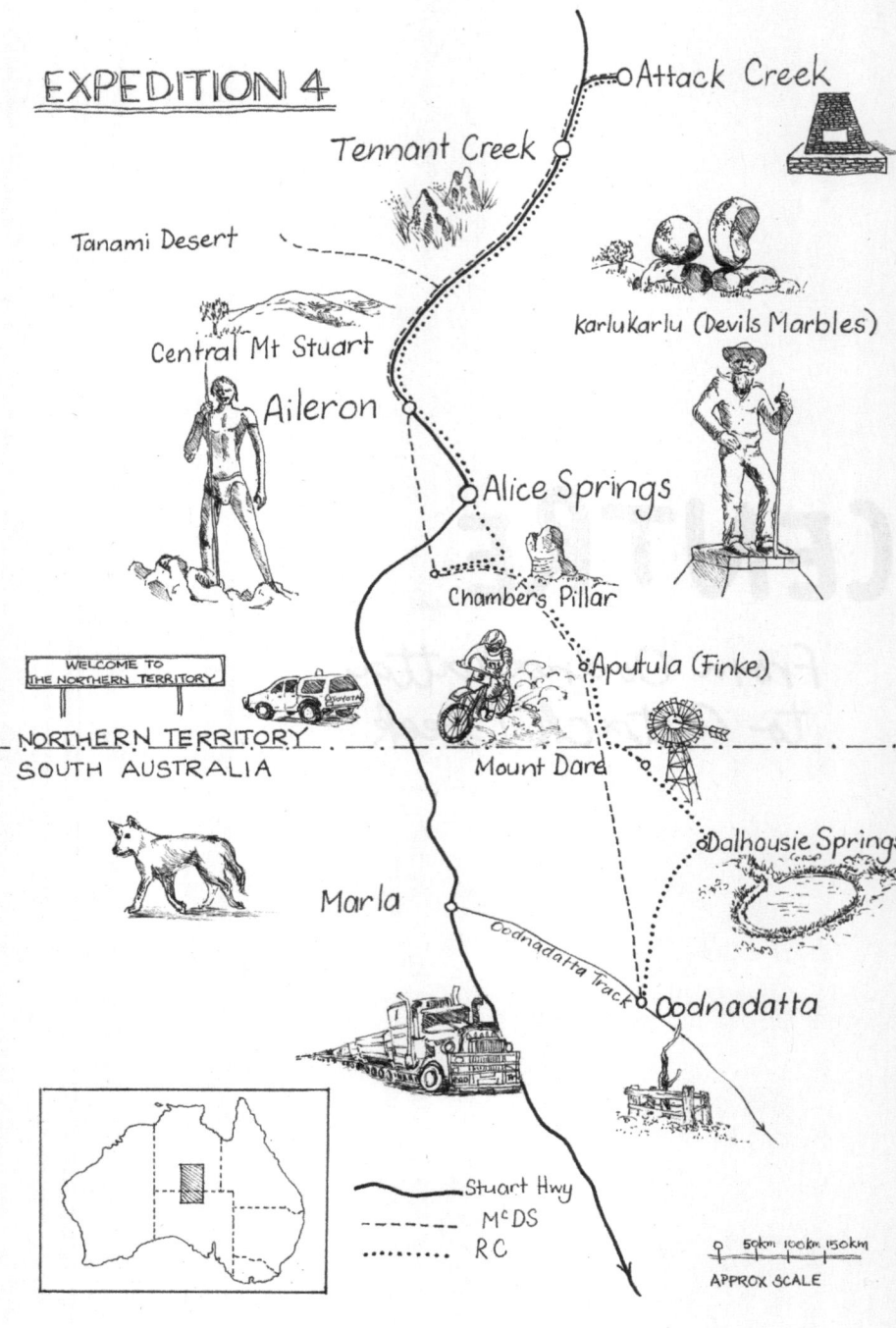

EXPEDITION 4

Attack Creek

Tennant Creek

Tanami Desert

karlukarlu (Devils Marbles)

Central Mt Stuart

Aileron

Alice Springs

Chambers Pillar

Aputula (Finke)

WELCOME TO
THE NORTHERN TERRITORY

NORTHERN TERRITORY
SOUTH AUSTRALIA

Mount Dare

Dalhousie Springs

Marla

Coodnadatta Track

Oodnadatta

Stuart Hwy
M^cDS
RC

50km 100km 150km
APPROX SCALE

SEVENTEEN

GLEAMING GIBBER STONES

Each time I unfolded and refolded my map of Central Australia—so many times it's in danger of ripping apart—I swore I'd no intention whatsoever, thank you very much, of heading due north from Oodnadatta into that expanse of nothingness.

Yet here I am, about to follow the most remote section of the Old Ghan railway, which followed the Overland Telegraph Line route constructed in the 1870s, which followed the routes taken by McDouall Stuart in 1860, 1861 and 1862. I blame those grainy black-and-white photos in the museum.

'It's a hundred and eighty kilometres to Dalhousie, so about three and a half hours. You'll be ready for that dip,' says Neville, handing me a mud map with directions. Dalhousie, within the Witjira National Park, is the biggest spring to spurt from the Great Artesian Basin.

'The rangers aren't out there at the moment, so no point in calling them if you get into trouble.' Neville points to the phone list.

'There's our number, and that's for the Mount Dare Hotel.'

Am I foolish? Driven? Maybe some would say sensible: I like to think so. After all, to keep going north from Oodnadatta to my next destination, Chambers Pillar, is a more direct route than detouring up the Stuart Highway and back south from Alice Springs.

Bad-Ass Betsie is up for it, even with a useless, stumpy aerial. And look at all the maps piling up on the passenger seat. All the directions I could ever need, while McDouall Stuart, maker of the first maps of the region, had nothing to guide him but his bush skills, following the flight of the birds and analysing the lie of the land.

'Petrol tank—refilled. Squirty water bottle—refilled. Food box (it's let me down so much I refuse to call it a fridge anymore)—refilled. I'm paying close attention to the government bulletin on driving in the desert, which seems to focus on wet-weather warnings. I look up at the cloudless blue sky and feel rather chirpy. It's amazing how refreshing it is to stop procrastinating.

McDouall Stuart had hoped to lead a decent-sized band of men on this, his fourth expedition. After all, he was planning, at long last, to be the first European to reach the heart of the continent and then the northern coastline, with hopes of mapping a route for the Overland Telegraph Line.

However, once again, he found himself the leader of a tiny team: himself, William Kekwick and Ben Head, a lump of a lad, who looked like he was used to being well-fed. Kekwick had done his best to get more men involved. In a letter to his brother, he admitted they had arrived back at Chambers Creek 'very much reduced'. Coupled with stories of McDouall Stuart's strict leadership style, this could account for the fact that Ben Head was the only stockman from Moolooloo both willing and allowed to join them.

McDouall Stuart was disappointed but chose to be cheery. After all, he had thirteen horses and lots more horseshoes than the last couple of times. They powered through the country, taking only three weeks to ride from Chambers Creek back to Oodnadatta. The terrain from here was unknown to any European. He must have softened towards the lad. The first chain of hills they came across, he named Head's Range, and the highest point he called Mount Ben. It's only 200 metres high, but out in this flat, flat land, that's quite a lofty landmark.

I come across a dead tree branch. If it wasn't for the shoogly fence surrounding it, I'd have driven right past. I read another of those pink and white hand-scribbled signs to jalouse that this is not your average tree limb. It's the Angle Pole. If this was once a telegraph pole, and a significant one at that, marking where the Overland Telegraph Line changed direction, it's a shadow of its former self.

I spot a road sign pointing left to the Stuart Highway. A final chance to change my mind. But I don't. Today, my map of choice is a ring-pull booklet of specific routes, now open at Witjara—Dalhousie Springs, with strands of multicoloured circles, like baubles thrown across the paper. I picture that blue and inviting expanse of water, and I'm looking forward to surprising McDouall Stuart, who failed to spot it on his expeditions.

Maybe that was understandable. McDouall Stuart contended with heavy rain, before and after Oodnadatta, creeks flowing too high to cross and country so boggy the horses were finding it hard to keep their feet. They lost one horse, but still, McDouall Stuart was chipper. Another horse threw a saddle bag and damaged the instruments and a hole discovered in another bag meant his folder of plans was soaked. Still chipper, though.

The page in the booklet I'm looking at gives descriptions of bogs I'll encounter driving in and out of Witjira National Park, each with dire warnings for wet weather conditions. Gluepot. The word sticks to the roof of your mouth. Tenacity. An obstinate word, a hiss of a don't-mess-with-me word. Great names. Here's another. Fogartys Claypan: when it's wet, the surface is soft and boggy. Challenging, it says, hazardous in the wrong weather conditions. The track through the dunes of the tiny Pedirka Desert to Witjira–Dalhousie Springs is inaccessible after rain.

Oh, how I laugh as I look up at the sky so blue and cloudless. Not a hint of rain. The creek beds I'm crossing are as dry as the bottom of that proverbial cocky cage.

The Oodnadatta Track had edges to follow but this road is a blur as it blends into the wide, red yonder. The road is red. The offshoot tracks are red. The land in between is red. I could drive off in any direction with nothing to stop me, go off the edge of the world for the hell of it. No fences, no boundaries. Freedom is what I'm feeling but I'm not sure if the racing of my heart is excitement or dread.

A jumbo-size post box for an isolated sheep and cattle station is ahead, signalling where I turn right along Public Access Route 8 to Witjira National Park. Things postal keep cropping up on this road trip, as though my postie dad has joined us. He did grow up in a wee hoose around the corner from McDouall Stuart's birthplace.

I can imagine the conversation between those two and I don't find it a comfortable one.

'The first word she spoke as a bairn was 'eelp' when she got stuck in a chair she'd pushed under the table.' That would be my dad. 'She'd been told not to. I don't think she's ever said 'help' again.'

'Aye. I've noticed.' That would be McDouall Stuart, finding any opportunity for a dig.

I'm on my belly taking photos of the gleaming gibber, cobble-stones blasted and polished over thousands, maybe millions, of years. Reminds me of the cobbled streets of Dysart, the fishing village where that pair grew up. I wouldn't like to ride a bike over either surface but Bad-Ass Betsie is doing fine, living up to the high expectations I had for her.

The drive to the Dalhousie campground is rocky, rough and slow. I should have paid less attention to the descriptions of those bogs in the ring-pull map and more to the multicoloured baubles. The red, orange and lemon dots and squiggles are code for sharp bends and dips, ruts and washouts, crests and holes. It's all spelled out in a list I'd ignored. I plough my way through the Pedirka Desert dunes, lurch across creeks—scarily deep but mercifully dry—and veer around rocky outcrops, bouncing over the higgledy-piggledy gibber. I'm concentrating so much on keeping upright that I don't notice I've been criss-crossing the Old Ghan railway.

Six hours it takes me to reach the campground where I discovered the one item I failed to refill at Oodnadatta was the gas canister for cooking. Wood fires are prohibited, so no hot dinner for me. Can't even boil a pot of water for a cuppa. Can't just pop along to the local hardware store to fix this problem. This is my lot until I get to Alice Springs. Three days away. I spot a public phone box and the thought of ringing home to whine for a while is tempting, but it takes phone cards only and I lost mine somewhere back in the Flinders Ranges.

A notice warns 'Dingoes are likely to cower in the presence of an aggressive human... but attack one who displays fear.' How about a human who displays frustration?

Another dingo notice points out that they have been known to nick off with your leather shoes. 'A dingo, grown accustomed to feasting on human leftovers (that could have been worded more carefully) may show aggressive behaviour if they cannot find any food in the wild.' I'm hoping no Dalhousie dingo has the munchies tonight. I can't imagine them being satisfied with the remains of my apple and carrot salad.

'You thought the water flowing out of the mounds along the Oodnadatta Track were springs... this is a spring!' I've been dying to say this to McDouall Stuart all day, though I'm sure he's unaware of the *Crocodile Dundee* reference. We're hanging our feet over the edge of the boardwalk stretching out over the largest in a supergroup of about sixty springs that extend over more than 50,000 hectares. The Witjira–Dalhousie cluster was first sighted by a small party of surveyors working on the Overland Telegraph Line in 1870. The Great Artesian Basin (GAB), the source of the water flowing from all the mound springs throughout Australia, was discovered a few years later.

The flow from the main spring is around 14 million litres every day. That sounds a lot, doesn't it? But we've never stopped tapping into the GAB like it was infinite and First Nations folk and conservationists around Australia are worried.

The Arabana native-title holders of the springs down near Kati Thanda Lake Eyre and the Wangkangurru and Lower Southern Arrernte people who have looked after the Witjira–Dalhousie Springs for many thousands of years, plus many other First Nations groups in Queensland and New South Wales, believe time is running short for the mound springs system and serious intervention is needed.

Colin Harris, a past president of the Royal Geographical Society of South Australia and President of Friends of Mound Springs volunteer group says much remains to be done and, from a national perspective, the situation is quite unsatisfactory.

In his 2017 John McDouall Stuart Memorial Address, Colin pointed to the incredibly rich First Nations and non-First Nations history and heritage involving the sweep of mound springs country, stating: 'The Indigenous song cycles and trading routes were on a continental scale, as were Stuart's journeys, the construction of the Overland Telegraph Line, the Central Australian Railway and the Stuart Highway.

'What is needed for this stretch is a national approach which involves the South Australian, Northern Territory and Commonwealth governments working with the Indigenous people of the region to conserve and celebrate a truly remarkable heritage legacy.'

EIGHTEEN

NOT YOUR AVERAGE SUNDAY JAUNT

I decide against a dip at dusk, promising myself I'd venture in at dawn. It's not so much the vision of a Nessie-type monster appearing on the shimmering surface that puts me off, nor the resident goby fish nibbling at my heels, it's the silent but ferocious mozzies.

It probably wasn't the best idea to check out dingo details before slipping into my swag for the night, fascinating though they are—their acute hearing, their keen eyesight, their strong sense of smell. Did you know they can open doors with their paws, just like humans, thanks to the ability to swivel their wrists? They can also turn their heads 180 degrees. We humans can only manage 70 degrees maximum.

As I listen to their far-off howl, it gets me thinking about Lindy Chamberlain whose two-month-old baby daughter, Azaria, was taken by a dingo in 1982 while the family was staying in a campground in the centre of Australia—not unlike the one I'm in now. Lindy wasn't believed and she was originally sentenced to life in prison for murder before being released after four years when new evidence came to light. It took her thirty years to be vindicated. The movie *A Cry in the Dark* starring Meryl Streep tells her story.

I paid attention to the mosquito warning, dousing myself with repellent, and the march flies are so big and dumb they're easily whacked, but something's bitten me on the belly button and I'm itching all over.

A rustle outside the tent, and I'm bolt upright. I think I may stay in this position all night long.

That invigorating dawn dip was anything but, with the temperature of the water soaring past 35 degrees. That would have sent me off to sleep last night. I'm as pink as a prawn after a very short swim and I'm drowsy, but at least I'm clean. I drive out of the park, the air-con on full blast and a wet towel around my neck doing a fair job of keeping me alert.

Perched at the bar in the Mount Dare Hotel, having covered eighty kilometres so far of the 350 planned for today, I could fool myself into thinking this is your average Sunday jaunt and this is a typical pub. Except I'm chatting to mine hosts Graham and Sandra Scott and their daughter Shaynee who are the entire population of Mount Dare. Rather than a carpark, there are two airstrips used for mail deliveries and by the Royal Flying Doctor Service.

The Simpson Desert, lapping at the front entrance, covers 170,000 square kilometres, crossing the corners of three states. The largest parallel dune system in the world with tracks rather than any defined roads, it is named after a man who made washing machines. You'd need a washing machine after crossing that expanse. The desert is closed in the height of summer 'to save people from themselves', says Graham. The Scotts have a rescue tow truck standing by and are involved more often than they'd like in the desert recovery of hapless travellers.

The Scotts are very much at home, so I'm surprised when they tell me it's only a few months since they swapped their cattle farm near the ski slopes of Mount Buller in Victoria to come here. We wander around the beer garden and the Scotts tick off an impressive list of visiting birds to the billabong: zebra finches, budgies, blue herons, corellas, galahs, wedge-tailed eagles. Once there was a pelican and one twitcher was delighted to spot a very rare Burke's parrot.

Never let a flushing toilet pass you by and fill up with fuel whenever you get a chance. I'm ready to leave. Standing at noon out the front of the hotel, Graham points in the direction of the border—ten kilometres away—and reckons it's four hours to Chambers Pillar. That means nearer five for me, still leaving plenty of time to settle in for the setting sun's performance, scheduled for 6.45 pm.

'I can take corrugations along the tracks, I'll ford a flowing river if I have to, but I avoid sand at all costs,' I tell them.

'I'll check your tyres,' says Graham and he reduces the pressure to around 25 p.s.i. Sandra hands me a slice of her fruitcake for the picnic I've planned for along the next stretch.

McDouall Stuart wrote a lot about trees around here, the mulga scrub so dense in parts 'it is far worse than guiding a vessel at sea; the compass requires constantly to be in hand'. Saltbush and spinifex are all I see breaking up the red, red landscape.

The sign I'm looking for looms ahead: Welcome to the Northern Territory. McDouall Stuart reached the 26th parallel on 2 April 1860 to become the first European to cross over the border into what was then a nameless land mass and part of the colony of New South Wales.

'How did you celebrate?'

'I had ither things tae think aboot. The horses for a start, the country wis cutting up their shoes no end. And ma eyes were in a bad state.'

'But you must've been excited...you must've thought about it. You got so close before.'

'I wisnae in the mood for celebrating.'

'How come you didn't mention it in your journal?'

'Like I said. I was seein twa suns instead o' one at that stage, ma eyes were that bad I made an error of a few miles.'

'Are you saying you crossed the border and didn't realise until afterwards?'

'Like I said, I had other things on ma mind.'

Well, *I'm* not going to let this moment pass. I raise my squirty water bottle to toast the border crossing.

With McDouall Stuart crossing over the border, South Australia's long-held hope of taking over this land all the way to the Indian Ocean was one step closer. McDouall Stuart had wanted the land to be called Alexandria. Kingsland was a popular suggestion to match up with Queensland. Another was Albert, after Queen Victoria's late husband. Centralia and Territoria were proposed. But 'Northern Territory' it was and has remained, shortened to the Territory or the NT.

Big cattle properties and big-brimmed hats: that was the vision for this land from the start, and it remains a huge part of the Territory's identity, often described as one of last of the world's frontiers. Maybe you think of a Wild West movie full of cowboys and cattle; the awesomeness of wide-open spaces; the smell of dust; the bellowing of the animals; the women on the front porch of the homestead in long white frocks wiping floury hands on their pinnies from baking bread. You'd be forgiven that view after the movie *Australia*. Or maybe you've been watching a different kind of colonial movie, something more fearsome, where the story is all about the lack of policing, the lack of control, the lawlessness.

Pastoralism was to be the Territory's economic saviour and, once Queen Victoria handed it over, the South Australian government issued leases for huge cattle stations, run like mini fiefdoms. Pastoralists fully expected to make their fortunes. By 1911, when the NT was handed over to the fledgling Commonwealth Government of Australia, there were 255 pastoral leases covering a quarter of the land mass. Today, pastoral stations cover half of the Territory. But the story is more bust than boom.

Cattle stations have to be big, as the land can't sustain more than a handful of cattle per square kilometre and the reality is a long list of issues: small local market, lack of infrastructure, stock diseases, and remoteness from markets with long distances to road or rail points for transport to ports and slaughter yards. Not to mention the loneliness, unreliable rainfall patterns and arid land.

Pastoralism has always been a case of heavy subsidies and support and, in the early years, reliance on cheap labour provided by First Nations families. They were the mainstay of the industry but were paid very little for their work, if at all. In the mid-1960s, 4,000 Aboriginal people were living and working on NT pastoral stations. Many saw it as a way of remaining in contact with their own country: the men mustering and yarding cattle, branding calves, fencing and butchering; the women working as cooks and cleaners; and the children given chores before and after school. Nowadays, aerial mustering and quad bikes have largely taken the place of stockmen on horseback and workforces are tiny.

Many pastoral leases were mismanaged, the land flogged and

ultimately abandoned with worn out and broken infrastructure. Since the 1970s, many unprofitable pastoral leases have been sold to Aboriginal entities and, despite the difficulty of making a profit from cattle, pastoralism remains an option in their land management and land use plans.

According to the Central Land Council, Aboriginal people on the land see cattle businesses as helping them achieve broader social goals like gaining financial literacy, keeping family together, generating pride and empowerment and maintaining culture and spiritual affiliation.

NINETEEN

WAGGING FINGER

Splat in the middle of an area with very few landmarks is Aputula (Finke), population of around 200, the next highlighted dot on my map. It looks helluva lonesome.

By the time I come across the fading pink lid of an oil drum nailed to a rusty stake, the towel around my neck is bone dry. The scrawl on the makeshift sign tells me I'm still sixty-one kilometres away. The CD has gone into repeat yet again when the lofty welcome sign to Aputula appears, dwarfing all the other signs along the white fence. They announce what's what in the Aboriginal community: a police sign, a first aid sign, speed sign, children crossing sign, no alcohol sign.

A distant buzz gets louder and closer and, revving through the fence opening, comes a youngster on his quadbike. He completes a donut in front of me, taking a good look before scooting off, back past that white fence.

Another, larger, sign tells the story about the railway bridge that was washed away one year after the Ghan railway line was opened in 1929 and how, over the next fifty years, low-level crossings, built and rebuilt, failed to handle massive floods. The line was shifted west in 1980.

I follow the youngster into the empty streets of Aputula. Once a year in June, this is where the two-day, off-road Finke Desert Race begins, one of the biggest annual sporting events in the Territory and advertised as the most fun you can have with a helmet on. The disused railway line runs parallel, sometimes merging, with the race track.

At the far end of the community, I take a punt on which way to go. Sand all around and three roads to choose from—one dead straight, one meandering by its side, the third wide but bumpy. The straight one makes sense for a railway track, but I choose the wide and bumpy one as it's more prominent. When it becomes sandy and almost peters out I lose confidence and drive back into the community, where a cheery lad tells me I was right after all. The straight road leads to the town's sewage works.

'Once you cross over the Finke River, you'll be on the railway track in no time.'

'Is there water in the river?' I'm thinking of that sign about floods.

'We wish,' the lad says.

I slide into the wide, deep—and dry—creek bed and out the other side, congratulating myself for keeping a firm grip on the steering wheel and my nerves. Seconds later, I spot a rail spike sticking up through the sand.

This stretch of the Ghan railway line was notorious for delays, sometimes for weeks, or even months. If it wasn't flash floods, it was sand drifts halting the journey. Delays were so common that an open rail car behind the engine carried extra sleepers and tools so the crew—and passengers—could repair the line.

Word of advice: a rail spike as a memento of the journey may seem like a good idea but this one had been sitting in the blazing noonday sun. I immediately flung it in the air like the red-hot poker it had become and nearly crowned myself. Driving over one of those massive nails would surely end in tears.

As Bad-Ass Betsie jolts and shudders over a patch of road that's turned into a giant washboard, that broken aerial stump flaps against the windscreen like the wagging finger of a strict schoolmarm saying 'I told you not to do it!'. Where is McDouall Stuart when I need someone to blame?

I plunge back into ruts of soft sand and Bad-Ass Betsie is slipping and sliding and ignoring my efforts to swap 'lanes' in search of firmer ground. She's turned into one of McDouall Stuart's skittish horses and I remember what Lyn Litchfield said back at Wilpoorinna. I breathe out and ease my grip on the steering wheel. 'Never stop feeling what the horse is feeling,' Lyn had said. 'You have to be totally focused.'

Betsie falters and threatens to grind to a halt in the talcum-fine bulldust as I change gears. The warning back in Oodnadatta slides into my head. 'Just don't let her stall, or you're done for.' Hands are clenched once more on the steering wheel, knuckles white, arms rigid like I'm physically pushing this two-tonne beast forward. I'm listening for the engine faltering. No music playing now.

It's a perverse relief when I'm back on bone-rattling, car-jostling solid ground. But that doesn't last. We dip once more into treacherous sandpits. It's a repetitive pattern. I glance at the clock. That must be wrong. Surely this torture has lasted for more than an hour.

'I've had enough,' I wail at full volume to the universe. 'I've. Had.

Enough.' The universe isn't listening.

The fuel gauge is slipping low, I'm itching all over but I can't scratch, and I hear a clink and I just know I've smashed the piccolo of sparkling wine I've coddled in a blanket since Adelaide, to open in celebration once I reach the Top End.

The road improves, the music is back on and I'm game enough to stop and take a photo of the ruins of Bundooma railway siding. I tune into the radio for the news at 5 pm. I'm pushing it but, with this fantastic road, I calculate I'll get to Chambers Pillar just in time.

It's 6 pm by the time I turn west, fill up with petrol at the general store at Titjikala (Maryvale) and begin the final forty-three kilometres to the pillar. Two kilometres beyond Titjikala, however, and it's back to the pattern of alternating corrugations and bull-dust—with the added touch of driving one-handed. The other is blocking out the blinding glare of the setting sun, floating ahead, like a big red balloon, beckoning me to keep pace, and promising to delay its evening performance over Chambers Pillar, just for me. I'm doing what I said I'd never do: drive at dusk. Along with kangaroos bouncing across the road, other potential animal hazards include feral horses, donkeys, dogs and the occasional camel.

Then I see it: a bloody great heifer in the middle of the road, looking at me with her big cow eyes. She isn't budging. The family joins her. My beep-beeps only attract more cow families out for their evening stroll. I hadn't factored in the unfenced cattle stations around here.

This has to be one of the stupidest things I've ever, ever done. Keep coming, keep coming, beckons that big red balloon and I have no choice but to plough on. There's no stopping and there's no going back. Any opportunity to call it a day has passed as the road has become too narrow for me to stop and camp. If I did and a car came along in the middle of the night...

The road rises and rises and rises. Bad-Ass Betsie, doing so well, twists around a hairpin bend at the summit. The view is spectacular. But it's not Chambers Pillar.

I weave my way down. On and on. Round and round. A gate announces entry into the Chambers Pillar Historical Reserve and a big black and orange sign brings me to a screaming halt.

'It's a scunner to get so close, but ye'd a been aff yer heid tae keep goin'. I have to agree with McDouall Stuart but I'm not happy.

'It's good to be canny.' That's his trademark style right enough.

Nearby is a fire pit ringed with smooth stones, conveniently made by a previous camper, who I imagine muttered, as I am now, 'so near and yet so far'.

I have a small window of light to see what I'm doing and, quick smart, I've set up the tent, relieved that the soggy sleeping bag is from a burst water carton and not from the piccolo of bubbly (miraculously still intact). After forty-plus years in Australia, I'm still astonished at how speedily the sun sets. It catches me out every time.

McDouall Stuart is eyeing my sleeping bag draped over the open car door to dry, and the sodden cardboard box sagging open over the deflated water bladder.

'How are you gaun for water?'

Fuck. Hadn't thought of that. No fuel for the stove and now running low on water. I'm taking this re-enactment of McDouall Stuart's minimalist expeditions a bit too literally.

The sky is showing off its sunset afterglow of brilliant red and orange and crimson and I crack open that piccolo. It's important to celebrate coming to my senses.

Warm bubbly and moist fruitcake. I've never had such a delicious supper.

TWENTY

LUCKY LITTLE DUNNART

The mozzies are gone, the morning fly shift hasn't yet arrived and I've managed to start a fire. The billy is boiling and, now that I'm on water rations, I've poured in exactly a cuppa's worth.

The sign that stopped me last night, with its big black letters and drawings of cars and steep inclines, looks cartoonish in the light of day. I've been awake for a while—never really fell asleep—spending anxious hours contemplating these final eight kilometres to Chambers Pillar and vowing to re-read the car's four-wheel-driving instructions. I don't have the flag recommended on that sign to warn oncoming traffic coming round the bends but I do have Bad-Ass Betsie's topknot mine light to flash. I also have a bone-dry piece of cardboard to help (perhaps) with any bogging incident.

Chambers Pillar has been on my bucket list from the planning stages for this trip, one of the few asterisks on that map on the wall back home. The tourist gumph says it's a spectacular solitary column towering fifty metres above sandy hummock grasslands and is most dramatic at sunrise and sunset.

I had flashed that light all the way to the campground but not a vehicle did I pass. For McDouall Stuart, this stretch was full of tricky moments. No road whatsoever for him to follow. He had trouble getting his horses over and around the high sandhills covered in spiky spinifex. The creeks, too, presented problems, one with a bed of quicksand, another taking three-quarters of an hour to get men and horses to the other side.

The geological explanation is that Chambers Pillar was formed 340 million years ago and that the red hard cap is due to weathering in more recent times—a mere 20 to 80 million years ago or so. The Aboriginal creation story is of Iterrkewarre, a fierce spirit ancestor who manifests as a knob-tailed gecko, a significant symbol for the Lower Southern Arrernte and Luritja nations.

Iterrkewarre was very evil, travelling widely and killing many. He lived with women who, under tribal law, were forbidden to him and he was banished for taking a wife from the wrong skin group. The pair retreated into the desert and when they stopped to rest, he was cast into sandstone, becoming this fifty-metre-high monolith. His companion Yayurara crouched down in shame and was turned into a nearby rocky formation which McDouall Stuart called Castle Rock. I too thought of 'castle' as soon as I saw that rock on the way in here. Other nationalities would perhaps see something else altogether.

Chambers Pillar is striking in every photo I've seen, and here I am, taking my own, ignoring the heat and glare of the noonday sun, forgetting my little water predicament. Pictures of the front and back, from near and far. I zoom in to capture the knobby bit on top, stand back to focus on the whole column including the mound below and then saunter further afield to get a landscape photo from a distance. Every photo is a winner. It's impossible to take a poor one.

I inspect the position in the dunes for sunset photography and the sunrise viewing area. If I don't stop taking photos, I'll have to hoon around the campground to recharge the camera battery for those shots. I am spoiled for choice for camping, eventually picking a spot with Chambers Pillar in full view on one side, and Castle Rock on the other.

I may be the only person here, but a wildlife poster points out that I'm sharing this campground with the usual suspects lurking in the Aussie undergrowth: venomous snakes (this time the mulga snake) and snippy little beasties (this time scorpions).

Apart from the slither of snakes and scorpions scurrying out of their burrows and sneaking into my bedding, some other night-time action I can look forward to is the swoop of the lesser long-eared bat passing over, and the hoot of the southern boobook. My nature stroll in the morning is to check around the shrubs and bushes for the spinifex hopping mouse. I wonder if they're the same as the ones that McDouall Stuart considered 'elegant and tasty' on his first expedition.

The stripe-faced dunnart, a petite marsupial only found in Australia, is around here too. It is so evolved to cope with the Red Centre that it never needs to drink, getting all the water it needs from its prey. Lucky little dunnart. There's not one tap around here to refill my water bottle.

Another billboard is about markings on the sandstone pillar. Pre-colonial engravings are a reminder of just how long the Arrernte people have been connected to this land. McDouall Stuart (who did carve the letter 'S' on a few tree trunks on his expeditions) didn't scratch his name on the pillar but plenty of people did: members of the Overland Telegraph Line survey team who arrived about a decade later, followed by pioneer pastoralists taking up land. They passed by, camped here, and left their signatures and initials, marking the arrival of colonisation to the region.

The Arrernte traditional owners managed to keep their connections with their land strong, despite conflicts with pastoralists. Many have memories of camping near Chambers Pillar and senior traditional owners remember gathering bush tucker and medicines, as well as hunting goanna and small wallaby. Today, no longer denied access to their land and with their traditional ownership recognised, they have substantial involvement in the management of the pillar and the surrounding reserve under a joint management plan with Territory government park rangers. They continue their customs, gathering bush foods such as bush tomatoes and witchetty grubs, bush medicines and pituri (bush tobacco) and passing on their knowledge of the local plants and animals.

'An unfortunate shape,' says Lorne, as we snap away at Chambers Pillar at sunset like it's some kind of competition for the number of photos we can take, his partner Anne egging us on. 'My mates will be asking me what I'm taking pictures of.' The couple had turned up late in the day to become my new best friends.

McDouall Stuart had described the monolith as 'a remarkable pillar which has the appearance at this distance of a locomotive engine with its funnel'. But the image now hinted at by Lorne is impossible to erase. The shape of landmarks in the landscape was often a starting point for Aboriginal Creation stories of a moral nature and my thoughts return to the Aboriginal story of the giant gecko punished for breaking important kinship laws that determined, for example, suitable marriage partners, everyday behaviour patterns, how people related to each other.

McDouall Stuart is said to have chosen this pillar to memorialise James Chambers, one of his major benefactors. However, John Bailey, author of *Mr Stuart's Track*, reckons it was Chambers himself who put his name to the pillar once he got his hands on McDouall Stuart's journal when he returned to Adelaide; perhaps because it sounded imposing? If McDouall Stuart had taken a camera, and if the morality tale had been known... Chambers might have chosen another landmark to bear his name.

I've been invited to join Lorne and Anne for dinner at their place, a very well-equipped rig that they've been driving around Australia for the past year. They're far too young to be grey nomads though.

We dine on mackerel that they'd caught off Nhulunbuy in East Arnhem Land a couple of months earlier. It's been travelling in their freezer for an occasion such as this. They steamed the mackerel with lemon and bush spice and served it with mashed potato, peas and corn. I supply the rice.

Our conversation turns to camping tips and tools. I do love a gadget and I produce my drink bottle with the trigger to mist your face. I'm a sucker for anything with a double purpose.

I mention that McDouall Stuart's second in command, Kekwick, sharpened the end of his spoon into a knife. Ahead of his time. I reckon he may also be the one who devised a way to fold over the handle on his pannikin for compact packing.

Lorne produces his own favourite multi-purpose tool—the end of a woman's stocking, which he proceeds to put over his face. 'Especially useful when using this,' he says, producing a chain-saw.

Good job I'd sussed out the sense of humour of this mountain of a man with a love of bird-watching, photography—and guns. 'Great for keeping out flies and dust,' he said through muffled nylon, 'when you're carving up a tree trunk that's fallen across the road.'

Before I head off to bed, Anne sprays WD-40 on the zips of my sneakers which have stuck halfway because of the sand and Lorne shows me their 'washing machine'—a big bucket with a tight lid that sploshes and washes your clothes as you drive. I'm going to buy both when I get to Alice Springs.

TWENTY-ONE

(UN)RELIABLE SOURCES

One hundred kilometres of unsealed road thump under the car bonnet, the bumps and dips nudging me forward. A final shove and I'm onto shiny, smooth bitumen.

The map of Central Australia on the passenger seat tells me I'm on the Stuart Highway—for the first time this trip—and it soon has me slicing through the MacDonnell Ranges at Heavitree Gap. Alice Springs, here I come. Too easy.

McDouall Stuart, on the other hand, couldn't have chosen a more difficult route through those ranges rising from the plains in parallel lines. He admitted as much in his journal. He led his two companions, William Kekwick and Ben Head, towards a bluff in the centre of a distant range, about seventy kilometres west of Alice Springs, pushing his way through dense mulga scrub and coaxing the horses through thick beds of dried-up, fallen branches. They passed through deep, narrow gorges, crossed wide valleys and trudged past towering cliffs, topped by large blocks of jagged rusty-red sandstone leaning against each other like toppling dominoes.

They were the first white men to set eyes on these ranges, 'the only real range I have met with since leaving the Flinders', noted McDouall Stuart. He named them after Governor MacDonnell of South Australia, who'd become quite a fan of this Scottish explorer's tenacity. The local Arrernte people, here for tens of thousands of years, call the ranges Tjoritja.

Four days it took the trio to trudge along a route now called Stuart's Pass before they reached the foot of what's now called Brinkley Bluff. Here is where McDouall Stuart listed their woes in his journal: their saddlebags were torn to pieces, ditto their clothes, and they arrived nearly naked. Polly, McDouall Stuart's favourite mare, was hobbling after being stabbed by a stick in the fetlock joint.

The white men spotted tracks all around but saw no one. Arrernte men also saw tracks, strange tracks of a four-legged beast they couldn't imagine existed and footprints of men with no toes. No wonder they were keeping their distance.

'Scurvy was a problem on every expedition, wasn't it?' I say.

'Aye. Ye've got that right, hen. A real menace. We did all we could on this expedition to avoid it. We ate pigface. We found anither plant that had fruit like cucumber and, at the spot we're talking aboot, the fruit of a palm tree had the taste o' a tattie once we roasted it in the fire.'

'You went a bit over the top in your description of that palm tree. You wrote of 'light-green fronds ten feet long, spreading out like the top of a grass tree. A remarkable tree.'

'Aye. I did that. It was fair brae to look at. Trouble was, the fruit wisnae fit to eat. Made the twa men violently sick. I hudnae eaten as much, so I wisnae affected.

'I cannae believe I climbed that bluff. The most difficult hill I've ever climbed. I had an even more terrible time gettin back doon again. One

false step and I should've been dashed tae pieces in the abyss below.'

From the top, he was rewarded with a panoramic view of the countryside, but he never got close to the site where the Alice Springs Telegraph Station was built a few years later. It was the only time his route and the telegraph line diverged significantly, as his passage through the ranges was considered too rugged.

'Bet you wish you'd seen this gap we're going through now, eh?'

'You're no wrong there, hen. It took us a whole day to move a mere five miles. I'm no surprised they looked for a better way for the telegraph line. I would've tae if I'd known.'

Alice Springs, splat in the middle of the desert, population around 30,000, is the biggest town in Central Australia. Such a pretty name, though the Lonely Planet reckons it would never win a beauty contest. I find a surprising amount of green in this oasis. The spiky spinifex that tormented McDouall Stuart has been largely replaced by buffel grass. That's the bright green blobs I see on the hillsides.

Arrernte Creation stories describe how the landscape surrounding Alice, including the MacDonnell Ranges, was formed by the caterpillar ancestors Ayepe-arenye, Ntyarlke and Utnerrengatye, creators of the water, trees and country of Alice Springs.

The Todd River snaking through the town, named after Charles Todd of Overland Telegraph Line fame, is mighty wide and dry. The story goes that if you see water in the Todd three times you've lived here long enough to almost be considered a local. The Arrernte people know this river as Lhere Mparntwe.

The river red gums lining the banks, hundreds of years old, are majestic, their gnarly girths too wide to properly hug. They've led a tough life; their roots forced to probe and prod through the parched soil to get water from deep in the ground. Annual rainfall is less than thirty centimetres (about eleven inches), and that's been known to fall in one day.

The town was called Stuart for a few years but that didn't catch on. It was changed to Alice Springs in the 1930s to match the telegraph station to avoid confusion. To those who've made this town their home, she's simply Alice, named after Todd's wife. The Arrernte nation calls it Mparntwe.

'It wis called Stuart, did ye say?' I nod.

'Well, I never saw that Todd River in full flow. Nor bone dry for that matter, so there's nae chance of me bein' called a local.'

'Alice Todd never saw the river either, if that makes you feel any better. She never got near the Northern Territory. She never left Adelaide.'

'Is that right noo.'

'There's no spring, by the way. Depression is the proper word for the water near where they built the telegraph station.'

'Alice Springs is a cheery name.'

'Could've been Stuart's Depression. Doesn't quite have the same ring about it.'

I go straight to one of the caravan parks and pitch my tent which looks small and lost among the motor homes and trailers. I head for a café along the road to check my notes about bush tucker and bush medicine and hopefully get a decent coffee. Despite the noonday heat, I've chosen an outside table in the garden. I'm not quite ready for walls. Am I turning into McDouall Stuart? Heaven forbid.

I'm loving this reintroduction to refined living; internet access, a comfy chair and a decent coffee, a fan wafting the air and an awning shielding the laptop screen from the sun's reflection. Sheer luxury. At this very moment, a room is being prepared for me at a posh hotel not far from here, where I've got three days' work at a remote health workers' conference. Fluffy pillows and a mattress off the ground. Will I want to return to basic camping after that?

One of my aims on this trip is to explore how much better McDouall Stuart might have fared if he'd been able to chat with the locals for tips on which plants to eat. Wonder if any of the organisations in Alice Springs I've emailed have answered. They haven't. Aboriginal people must get jack of journalists and writers asking the same old, same old questions, over and over. They can't be bothered answering, perhaps, or they're too busy.

'Looks like you're busy.' I look up at a man indicating the spare chair opposite me. I notice his clean fingernails. 'Mind if I sit here?'

'Taking advantage of the wi-fi,' I say, curling my grubby fingers over the keyboard, hopefully establishing that I am a traveller passing through to excuse the dirt under the nails. I've decided he's a local.

'I'm not a local.' He's reading my mind. 'Only been here for a couple of years.'

'And never seen the Todd River in full flow,' we say in unison. We laugh. This could be a fun encounter and I let myself daydream about a date. Someone to take me out on the town tonight. It's been a while.

I launch into my story. I've got used to folk I meet being interested in McDouall Stuart. The pastoralist who's built a statue of him on his property, the publican who's created a museum at the back of his pub, the postie who tells travellers about the explorer's exploits. They're impressed with his bush skills and achievements, yes, but even more with his persistence in pushing through. Exploring this country once is one thing; coming back again and again is another.

'Don't expect a warm reception here,' says Café Man. A reaction I'm not expecting.

He told me he'd been to a meeting a few years ago in Alice Springs protesting about the arrival of a statue of McDouall Stuart. What was it all about?

Café Man mentions a leaflet, slipped behind people's windscreen wipers during a protest rally, saying McDouall Stuart's journals reveal he was involved in massacres. What? I've read those journals. All six of them. Did I miss something?

Café Man reckons I'm following the tracks of a mass murderer. Not someone you would choose to spend time with on the open road. He's got it all wrong. McDouall Stuart was keen to avoid conflict. I agreed with those scribes who said that aggression wouldn't be in his best interests. Considering he'd be coming back down the same way, that seems logical. It makes sense, but have I been biased and cherry-picked the information I wanted to find? Maybe I'm the one who's got it all wrong.

Café Man says his information is from reliable sources. I tell him my sources are reliable too. A headache threatens as I dredge up memories of what I'd researched about Attack Creek near Tennant Creek, 600 kilometres up the Stuart Highway. McDouall Stuart named it that because he was attacked... wasn't he? My reading was that he hoped the band of warriors at Attack Creek would back off, but it was he and his two companions, weak, tired and hungry, who were forced to retreat. McDouall Stuart 'justly earned a reputation for humane and careful interactions with Aboriginal people'. That's Dr Philip Jones, curator at the SA Museum. 'Stuart's friendly attitude to the natives is reminiscent of Sturt.' That's Mona Stuart Webster in her biography of her great-grand-uncle, but she would say that, wouldn't she. She's family. I'm McDouall

Stuart's defence lawyer one minute, prosecutor the next. We're not related, but we do come from the same place in Scotland. Maybe I'm being protective?

I press Café Man. Who is a contact I can talk to in Alice Springs, I ask, and he says he'd have to check that out. I give him my phone number, not for the reason I'd briefly entertained earlier. That went well and I don't think. It's unsettling.

I get back to my list of contacts. Instead of securing interviews to chat about bush tucker, I'm coming up against energetic gatekeepers who are slamming their gates shut. I discover the reluctance to talk to me is because of McDouall Stuart himself. 'There are Aboriginal women in our organisation who have the knowledge, but doubt they'd want to talk to you.' 'Not a lot of sympathy for McDouall Stuart in these parts.' 'People around here aren't keen on explorers.'

I briefly consider not mentioning McDouall Stuart's name at all, but discount that notion immediately. Maybe I could mention his name and say I'm no apologist for him. Nah. Mealy-mouthed.

Ping goes the laptop and a message pops up. 'Don't think they'll want to speak to you.' I'm tempted to go back to the caravan park and finish off that wine cask.

TWENTY-TWO

SAGA OF THE STATUE

'Fine company you're turning out to be,' I'm muttering as I drive towards the town centre the next morning. 'Pisspot is one thing, but mass murderer...' The air-con thrums, the engine whirrs, but no response from McDouall Stuart.

One in five locals in Alice Springs is Aboriginal. It seems a lot more, maybe because they're the only people I pass as they walk from the town camps dotted around the fringes of town. Wherever I look, as I saunter along the shopping mall, Aboriginal culture is up for sale. Aboriginal artists, with one canvas or more laid out on the grassy verges, try their luck to get the attention of passers-by, while art galleries display racks of canvases ready for framing. Alice Springs is said to have the most art galleries of any town or city in Australia.

Tourists pause at billboards offering all things Aboriginal: guided visits to communities, bush tucker experiences, cultural walks around town and four-wheel-drive tours to rock carving sites.

Those on a budget for their memento from the Territory might be drawn to the mass-produced boomerangs in the gift shops.

In 2000, the Arrernte people were acknowledged as the traditional owners of Alice Springs, becoming Australia's first First Nations group to get Native Title rights in an urban area. It was no mean feat for them to prove their continuous link with this land, considering that, between 1928 and 1965, this town was out of bounds for all Aboriginal people.

I head for the library at the civic centre and see Aboriginal mums sitting on the lawns, trying to hand sandwiches and drinks to kids more intent on playing with their mates. Sitting outside, waiting for the library to open, I watch elderly Aboriginal men in stockmen's trademark wide-brimmed hats and check shirts milling on the footpath. Alice Springs has become the service centre for about 260 remote communities in the surrounding region, roughly the size of France, and I listen to those old men speaking many languages—Arrernte, Anmatyerre, Warlpiri, Luritja, Pitjantjatjara, Pintupi—as they board minibuses to transport them back home.

The non-Aboriginal population is quite a mixture too: local families who are direct descendants of early settlers with a long history and a deep connection with the region, and public servants, university bods and bank workers here for a short or a long time. Some came in the 70s to protest the Space Research Facility at Pine Gap and stayed, and others came to work there, with 2,000 Alice Springs residents today holding American citizenship. Then there are the blow-ins—Australian and international workers on short-term contracts.

The promotion of Alice Springs as a 'tourism mecca' focused on all things Aboriginal and its label as 'an Aboriginal ghetto' in media commentaries is the contrast laid bare in an award-winning essay by local author, journalist and university lecturer Glenn Morrison. He describes Alice Springs as 'desperately trying to heal its own divide', a sentiment mirrored in an Australian Institute of Criminology report which notes 'a lack of general interaction between Indigenous and non-Indigenous residents.'

A poster informs me that the Gathering Garden in front of the library strives 'to strike a balance and create connections between the region's ancient past and more recent history'. It includes a collaboration with First Nations artists from various communities surrounding Alice Springs and engraved bronze coolamons (Aboriginal carrying dishes) flipped over to provide seating. A collection of plaques and notice boards gives a nod to the early pastoralists, Afghan cameleers, miners and various local identities. At the library entrance, there's even an installation of giant books to the memory of Neville Shute, author of *A Town Like Alice*. The statue of McDouall Stuart arrived on these lawns in 2010 to add another fragment to the region's story. It was a gift from the Freemasons, as McDouall Stuart had joined the organisation's Lodge of Truth in Adelaide between expeditions.

The statue was crafted by local folk sculptor Mark Egan, who is also the creator of Anmatjere Man, which stands at the Aileron Roadhouse on the Stuart Highway, 130 kilometres north of Alice Springs. Like the giant Anmatjere Man, the statue of McDouall Stuart is made of ferro-cement over a steel frame. It is four metres high on a one-metre-high plinth. While the Anmatjere warrior, several times taller, is pointing a spear in the air, McDouall Stuart is grasping the barrel of a rifle, the butt resting on the ground. Weighing in at two tonnes, McDouall Stuart's statue would be hard to miss. But it's nowhere to be seen.

The saga of the statue unfolds in the library as I sift through a motherlode of online and print media. The statue's arrival on the lawns was to be the culmination of celebrations, a series of talks and walks to commemorate the 150th anniversary of McDouall Stuart reaching the centre of Australia.

However, a public meeting the day before the unveiling raised questions about the insensitivity of the gift in a town sensitive to racial tensions. That was the meeting Café Man spoke to me about.

The council backed down but held a face-saving unveiling anyway. Visitors had come from Adelaide after all. Speeches were made, songs were sung and cheers mingled with calls of 'shame'. The bagpipe players eventually ran out of breath, the statue was taken down, and it languished on sculptor Mark Egan's property on the outskirts of town for four years. Some gift that statue was turning out to be.

In 2014, the statue was erected again, with much less fanfare, this time in Stuart Park, a heritage precinct a block away from the civic centre. The council had untangled the bureaucratic problems around who gave permission to accept the gift, who was going to pay for its erection and where it should be put. If the council thought this was going to pass unnoticed and unremarked, it was wrong.

Within days, a rally was held, calling for the statue to be removed. Arrernte elder Leonie Kngwarreye Palmer invited people to come forward and immerse themselves in the smoke. 'Smoking is really good for healing,' she is quoted as saying. 'For what happened in the past, we're just going to smoke it away and live for our future.' Meanwhile, that flyer with the heading 'McDouall Stuart is a murdering racist' was slipped under car windscreen wipers.

One year later, a crane turned up and hoisted McDouall Stuart on high, possibly giving protesters false hope. However, this third move—more a pirouette—was for safety reasons. Tourists were walking into the middle of the road to take photos.

Going by the screeds of articles, opinion pieces and cartoons I trawled through in the library, Alice could cause an argument in an empty house. If she were sitting in the therapist's chair, she'd reveal multiple personalities that sometimes rub along but generally don't, each blaming the other as to why.

The optimist in her felt the statue could start a conversation the town had to have, her pessimistic persona sensed a threat to split the town apart, while the wag wrote: never has one statue in the history of Alice Springs created so much verbal effluence.

I make the mistake of entering the swampy depths of letters to the editor columns and online comment threads exposing Alice's splintered identity. The focus of comments ranges from the relevance of McDouall Stuart, the statue's design and the debate about colonial-era memorials in general. Then it gets personal with strong opinions about McDouall Stuart, eventually accusing him of being a murdering racist, and strong opinions between townsfolk, accusing each other of being patronising, parasitic do-gooders and a bunch of control freaks.

The complaint about not being listened to, not being heard, not being consulted is a strong thread from all sides.

I visit the statue that took more years to find a resting place than it took the man himself to complete his six expeditions. I look up. And up. McDouall Stuart looks positively lanky as he leans on his rifle, which is also stretched long and skinny, a symbol of oppression to some, part and parcel of an early explorer's apparel to others. For a town council loath to pay for one installation of the statue, it has now forked out for three moves.

Thanks to the final shift, the pirouette, McDouall Stuart's gaze is no longer towards his beloved outback but southwards to Adelaide, the city he kept leaving. His broad shoulders are slightly hunched like he's trying to diminish his height and wishing he was anywhere but here. Anyone travelling along Stuart Terrace is now granted a view of explorer Stuart's backside, as a comment in a local newspaper observed.

A couple of Aboriginal kids come over to the statue to take turns running up the plinth and touching their statuesque playmate before twirling and jumping onto the soft grass. They are oblivious to suspicions that their launching pad, shaped like a pyramid, albeit with its top chopped off, is furtively promoting Freemasonry.

I catch a movement out of the side of my eye and notice for the first time a group of mums and aunties, sitting in the shade under yonder tree, keeping an eye on the kids. A second family gathering, a discreet distance away, also blends into the surroundings. It's like a painting revealing new layers to those who are patient but offers no hint of the statue's backstory: tortuous for the local council, disappointing for the sculptor, a flashpoint in the community.

A solitary bloke sitting under a tree waves me over. Jutting his chin in the direction of my car with its South Australian registration plates, he aims for a connection, telling me he's from Ernabella, a community in the Anangu Pitjantjatjara Yankunytjatjara (APY) Lands over the border. As fellow South Australians, we share my sandwich and then walk over to the statue. I ask if he knows anything about McDouall Stuart. Tall fella, he laughs. He was a short-arse in real life, I say. As we walk away, a handful of men take their place along McDouall Stuart's narrow strip of shade.

My camera is full of photos of the statue but I could not get one that brought McDouall Stuart out of the shadows.

it will be unnecessarily hurtful and provocative to many of the indigenous residents of Central Australia

the statue is offensive

Patronising and parasitic do-gooders

a total load of bollocks

Stuart is part of the history of whitefella Central Australia and should be celebrated accordingly

the statue will be inflammatory

an already troubled town

indigenous people do not need to be reminded of the violent nature in which their land was invaded and seized by our colonial settlers

don't tell us what to think and feel

simply inane

unnecessarily hurtful

it is a great shame job in this time when everyone is talking about reconciliation for our council to erect this statue that denies the existence of Arrernte people and to place it in the heart of Arrernte country

if you're going to live off the land you need a gun

that rifle is a celebration of genocide

the depiction of Stuart holding a gun scares Aboriginal people and is not what we want to be reminded about

the statue is offensive to me as an Australian committed to forging a future that respects all cultures that make up our nation

I think the statue was a wonderful gift to our town, and due to the nature of the subject matter, naturally, it will put some people off

what twaddle

175

It's the usual campground mix that evening: a pair of teachers on long-service leave, a fly-in fly-out worker heading home, retired couples. We do what you do in campgrounds: share travel stories, offer camping advice and generally wonder what the poor folk chained to an office desk are planning for their precious weekend off work.

'You can't win—whatever you write.' 'Not worth the hassle'. I'd mentioned my earlier encounter with Café Man to fellow campers around the communal open-air kitchen. 'You'll be called a racist.'

'Stick to bush tucker,' one of the teachers says. 'Stay away from Aboriginal issues. You'll be deliberately misunderstood by one side or the other, and probably both.' I've no intention of staying away from Aboriginal issues. Remote and rural health workers from all over Australia are attending the conference where I'll be writing stories about the proceedings. That gathering won't flinch from talking about inequalities in Aboriginal communities. The presenters know their stuff about health conditions that persist in remote areas of Australia, diseases that shouldn't be tolerated in a first-world country but are. I'll ask about scurvy.

'Ah. The Aboriginal problem.' 'What's to be done.' 'No one has the answer.' Now we're onto responses that usually stop any further debate, followed by a communal helpless shaking of the head. I point out that the participants in the conference will be proposing all manner of solutions, from pathways for Aboriginal people who aspire to study health at university to programmes within communities, such as caring for country, all of which could have positive effects on health outcomes. But I've lost their attention. The teachers are more interested in my hotel room. 'We'll take it off your hands,' says Jean, lusting after the mattress off the ground.

'Plump, fluffy pillows,' sighs Sue.

I lie in my tent. I never get bored looking at the stars. *Yer gettin' mair like me every day* is what McDouall Stuart might say, but he seems to have made himself scarce. I'd deny that suggestion if he was around.

TWENTY-THREE

TELLING TALES

The fluffy pillows are tempting but they'll have to wait. I'm talking on the phone to John Bailey, author of the biography *Mr Stuart's Track*, the book used to 'verify' the quotes on that flyer. He wasn't hard to find.

'I never suggested that there was a massacre,' he tells me. 'And no credible source, in fact no source at all that I know of, suggests there was. Certainly not my book. I would never say Stuart was deliberately violent. He was not a bloodthirsty man.'

He hadn't been contacted by anyone connected with the protest rally or the flyer. Neither was he contacted by any media outlet that covered the story, he said.

'What I was implying, assuming, or gathering from the evidence,' he continues, 'was an opinion—nothing more than an opinion— that it was highly likely some people were shot and highly likely that some were killed on his expeditions. In those days, if you had a bad wound, survival was not guaranteed.' He points out that recent historical research makes it clear that the frontier wars throughout Australia were real and should be acknowledged.

I also track down Stuart Traynor, author of *Alice Springs: From Singing Wire to Iconic Outback Town*. He tells me: 'McDouall Stuart had a situation there but there's no evidence that he had an overly aggressive attitude or that he killed a large number of Aboriginal people. I've read his journals over and over and I don't see any evidence of that.'

Dick Kimber, another author who lives still in Alice Springs, has been quoted as saying it was inconceivable that McDouall Stuart did not shoot anyone. 'Given that Stuart had always tried to avoid conflict, he must have felt he had no other option,' Kimber has said. 'I believe they considered they were in too much danger for warning shots.'

'I think that's the case,' said a volunteer with the National Trust in Alice Springs, 'that when he and his men were attacked, they attacked back. It's fairly general knowledge that he or some of his men would have killed because they were threatened.'

My double standards have been showing. I'm well aware that conflicts arose as colonisation crept through the country, the original inhabitants fighting for their land and the newcomers fighting to take it over. I'd wanted to believe there was no bloodshed on McDouall Stuart's expeditions.

During the health workers' conference, my failed attempts to speak to Aboriginal people about bush tucker come to mind as I listen to a presentation written by Pat Anderson, Alyawarre woman, founding chair of the Lowitja Institute. First Nations people find they are not spoken to, rather they are spoken about, she pointed out.

'It's a tragedy that racism still exists in Australia,' is the message in the address from CRANAplus CEO at the time, Chris Cliffe. 'The evidence is clear: racism has a direct and profound impact on not only the emotional but also the physical health of victims.'

'Our nation is a wealthy, well-connected, peaceful, developed country,' he said. 'But still, the further away people live from a capital city, the worse their health becomes.' This is because of the inequity of services in remote areas compared to the cities.

And a call to action comes from Professor Roianne West, born and raised Kalkadoon on her grandmother's ancestral lands in North West Queensland, who outlines her vision.

'I want to see my people socially and economically included in the nation. I want to see our kids educated, healthy and with the opportunity to grow up and get jobs and to be engaged in meaningful activity. I want to see our people living in good housing. I want our children to have good food to eat, fresh air to breathe and clean water to drink.

'Not too much to ask.'

Missionary, Mercenary or Misfit ('the 3Ms'), the categories that white folk working for Aboriginal organisations can fall into, is another presentation at the conference. It slaps me in the face. The labels describe today's do-gooders, whose mission is to save and rescue their clients; people who are in it for the money; and people who are hiding from themselves or something else. The presentation abstract suggests those labels may be becoming less relevant but the ABC TV series *8MMM Aboriginal Radio,* Australia's first comedy television series based in Alice Springs, suggests otherwise. Written, directed and performed by First Nations people, they poke fun at the misguided 3Ms working at the station. (I have a sneaking suspicion I've displayed 3M tendencies at various times in my career working for Aboriginal organisations.)

I'm packing up at the end of the conference when there's another ping on the laptop. I've got an interview to talk about native plants and bush tucker. Not with an Aboriginal person, but as close as I'm going to get on this trip. I'm meeting Peter Latz, retired Senior Botanist at the Parks and Wildlife Commission of the Northern Territory, considered the foremost non-Aboriginal authority on desert plants. Peter is the son of Lutheran missionaries and grew up in Ntaria (Hermannsburg Mission) west of Alice Springs, where he learned the Arrernte language and how to live off the land.

'I grew up with the Arrernte and heard lots of stories,' Peter says when we meet in the Bean Tree Café at the Olive Pink Botanic Garden.

McDouall Stuart's appearance at Brinkley Bluff west of Alice Springs may have produced the Arrernte word for horse, 'nanthua', says Peter. 'My guess is one of the locals had spotted the tracks of the horses on that expedition and reported back to the group, who refused to believe him. 'No, no, impossible,' they said, 'there's no animal like that,' so he led them to the spot and pointed to the evidence. 'Nanthua' the spotter said, which means 'here it is' and that's been the Arrernte name for horse ever since.'

When the Arrernte men finally tracked down McDouall Stuart's group, they decided they were looking at huge, two-headed animals, until the men dismounted, Peter suggests. 'And they were even more scared. These pale-skinned human-looking beings were possibly their ancestors, come back as spirits.'

But they soon realised this couldn't be the case when they saw them eating the seed of the cycad, an ancient plant that has a poisonous rind, probably evolved to stop the dinosaurs from eating it. 'The locals would know McDouall Stuart was less than godlike', says Peter. 'No god would prepare the cycad seed for eating by just roasting it.'

There are indeed many plants they could have eaten if they'd time to work with the locals but McDouall Stuart was very regimented with 'a strict regime of, up at sunrise, you do this and do that and keep going,' says Peter. 'He didn't have time to build relationships.'

And Peter suggests that I don't have enough time either, when I tell him of my wish to get an opinion about McDouall Stuart from an Arrernte person, based on stories handed down through generations. 'To get the knowledge would take you months getting their confidence,' he said.

And so I leave Alice Springs. McDouall Stuart's route from here more or less follows the Stuart Highway. I could scoot right up in a couple of days and still say I'd completed my goal to get to the top.

TWENTY-FOUR

WHO DO YOU THINK YOU ARE?

Road trains, kangaroos and grey nomads. The three dangers for drivers on the Stuart Highway. The motto for all three: Keep Your Distance.

You'd think the Stuart Highway would be a multiple-lane affair, befitting Australia's first transcontinental highway and the principal north-south route through the centre of the continent. Turns out it's single file each way, with few offers of a hard shoulder for emergencies. Slide over the fraying edge of the bitumen and you could end up with more than a gravel rash.

'The Track' or 'The Bitumen' are the names given by folk who use it regularly. Exploring not being their top priority, they know little and care even less about the man who gave the highway its name. They're too busy transporting freight across the continent or just want to get where they're going. As fast as possible. I keep stopping and pulling to the side to let them pass. Bad-Ass Betsie seemed so big, drawing attention to herself, on the streets of Adelaide. She was just the right look for the Oodnadatta Track and the Old Ghan railway, your reliable and capable outback lass who could go where others—too puny or too precious—feared to tread. This is different. I'm intimidated.

Another road train screams up behind me, a hundred tonnes-plus of heavy metal filling the rear vision mirror, headlights blazing, horn blaring. A trembling window separates me from the towering wheels hurtling past. One trailer, two, then a third, a never-ending blur of circling fury sucking me sideways into a vacuum. I cling to the steering wheel. Keep in a straight line, I beg my Betsie. Another prime mover looms from the opposite direction, towing a couple of trailers, the turbulence leaving me and Betsie shaking and shuddering. These monsters, the length of an Olympic swimming pool, stop for no one. They have so many wheels, if one blows out, they keep on keeping on. That explains the ribbons of rubber strewn along the highway.

I spy what looks like a black bedsheet flapping in the distance. As I approach, the sheet rises into the air, splitting apart, turning into a couple of wedge-tailed eagles that have been scavenging the bloody carcass of a kangaroo lying across the road.

Kangaroos are in their millions in the Northern Territory, far outnumbering humans, who amount to around 250,000. While Skippy looks cute bounding across a paddock, he is not so attractive if his ninety-kilogram body bounces onto your car bonnet. As a lump of roadkill, he's an accident waiting to happen to you. Bad-Ass Betsie and I were lucky to keep our distance—this time—from Danger Number Two, but I'm guessing it won't be the last encounter.

Not far out of Alice Springs, Bad-Ass Betsie and I veer off the Stuart

Highway where the map shows a cross-hatching of minor roads.

'You know yer goin in the wrong direction—I went west.' This is the first word out of McDouall Stuart since we first approached Alice Springs.

'I know,' says I. 'I'm not ready for the fast lane.'

'Aye. For me, it's big toons. That's why I left you tae yersel' back there.'

It's not just the driving that's stop-start as I thread my way through cattle country, jumping out to open and close gates, slowing down for meandering mobs moving to the next green patch to chew the cud.

Our conversation, which stuttered into action, again stops altogether. This is awkward. I'm not telling him what I'm thinking. I've a feeling he's keeping his thoughts to himself, too.

I cross over Plenty Highway and then cross over Sandover Highway, both stretching into Queensland. They are considered main roads, though they're mostly unsealed. I'm focused on following the dashed lines listed for four-wheel drive vehicles only on my map. The warning note in the bottom corner of the map was right about washouts, sand, bulldust patches and extreme corrugations giving me a good shake. A good shake is what I need. No mud is a blessing, but when did it last rain here?

Prickles are a pain, sneaking inside my socks every time I crunch over dust-dry grasses to get the perfect shot of a lonesome acacia. The top twigs of one long-dead specimen are wrapped in delicate barbed wire like they've been twirled through a macabre candy floss machine. I can only think it's the work of a willy-willy kicking up a dust storm. I speed downhill on a roller-coaster slope and glide up the other side in a whoosh where a tree, a live one this time, spreads its welcoming branches, saying, photograph me, photograph me. Oh, all right. Just one more.

The first vehicle in an hour passes by as I open another gate and some lads in an open traytop wave a thank you. A signpost to Gemtree Caravan Park beckons. Its website offers an escape from the hustle and bustle and that's exactly what I get. If it had been earlier in the year, I could have entered a paddy melon bowling competition, watched a movie, gone on a fossicking tour, or had a roast dinner served at the camp oven kitchen. I'm more than satisfied with an aimless wander around the station, taking close-up photos of flowers that are probably weeds and the first signs of termite mounds, though here they're not towering. The termites have slap-dashed their claggy soil over fallen trunks. There's a rustle and a kangaroo, hidden until I disturb him, hops away from me towards the sun dipping down, the tips of his fur glowing reddish-brown. The golden hour.

Who needs fluffy towels in a five-star hotel room?

Celia, a government relief worker and my only companion in the campground is horrified when I admit all I want to do on Stuart Highway is pull over and cower when a road train appears. 'God, no. Keep your nerve and keep up the speed,' she says.

'Think about it,' and she begins to talk very precisely as if I might be a bit slow on the uptake.

'If you slow down, the truckie coming from behind has to slow down too.'

She also tells me to make sure I have a clear path of more than 1.5 kilometres on the other side of the road should I want to overtake one of those metal monsters. That advice was unnecessary.

I face the truth and raise my cup of tea. 'My name's Rosemary, and I'm a grey nomad.' Danger Number Three. That's me with my Bad-Ass Betsie, a name that she totally owns: part profane, part twee;

the perfect name for a grey nomad's vehicle. Tens of thousands of grey nomads—of us—are out there at any one time, taking our time. Too much time for some. The legal limit for cars on some sections of the Stuart Highway is 130 kilometres an hour, the highest in the country. Not so long ago, the Northern Territory government played around with no limit at all. If I get to 100 kilometres an hour, I think I'm doing well. 'Travelling baby boomers' is another title we get; in the United States, it's the rather romantic-sounding 'snowbird'; here in Australia it's 'old farts.'

I have a bigger truth to face than being a slow coach. As a migrant I've sometimes felt like an outsider and, while probing Australia's colonial history, a bit of an onlooker. But the goalposts have moved. That statue back in Alice Springs has got me probing my relationship with this country I now call home. In the citizenship ceremony, you accept your rights and your responsibilities as an Australian citizen. I'm embracing the right to question, to criticise as well as praise what is done in our name and, if asked, 'Who do you think you are?', I can answer with a firmer footing. To acknowledge our colonial history, I now recognise, is a responsibility for everyone. McDouall Stuart was part of the settler colonisation process in South Australia, as I am now.

In McDouall Stuart's day, colonising nations around the world saw themselves as bringing civilisation to barbaric and savage nations. When I came to Australia, that paternalistic attitude and the sense of superiority and entitlement wasn't far below the surface. If you did mention colonisation, it was okay if you associated it with modernity, progress, and development. Anything else was un-Australian.

Settler colonisation, what we have in Australia, goes further than an attitude of entitlement to plunder the land and its resources and use the inhabitants as cheap labour. The powers that be were intent on establishing themselves as the new and rightful inhabitants of Australia. That First Nations people would soon die out was not just a belief, it was the plan.

Shunting the original inhabitants onto missions and denying them access to their country exposes the reality of a system geared to take over completely. Same with the policies that tried to erase language and culture, that legitimised taking children away from their parents, that controlled their every move, who they could associate with, where they could go.

And the effects linger on.

Did McDouall Stuart ever question himself, what he was doing? His motives for doing all this exploration?

'I've heard of him,' says Celia, when I tell her I've been following McDouall Stuart's route through the centre of the continent. 'I like the way he went about things,' she says.' 'Burke and Wills overstretched themselves, while Stuart, he knew when to stop and when to keep going.'

Do I know when to stop, when to keep going?

TWENTY-FIVE

THE DAWN OF LIBERTY

I sidle back onto the Stuart Highway, satisfied with my 150-kilo-metre detour of gates, grids, dirt and dust.

Dots on the map suggest I'll pass through little townships as I drive up the Stuart Highway but it turns out they indicate not much more than a roadhouse. A Hollywood-style sign on the hill announces that the first dot, Aileron, is ahead. It's a roadhouse, a smattering of houses, plus Anmatjere Man rising twelve metres— seventeen metres if you count his spear.

The sculptor Mark Egan, who also designed and created the McDouall Stuart statue in Alice Springs, said he thought for years about sculpting this statue, inspired during his years growing up in the Outback, seeing people living their traditional way of life. They were, he says 'just doing what they do best, and they're just amazing'. Maybe his statue will get passers-by to think a little, he said, while also giving them 'something to brighten up their day.' The local Anmatjere people, ever hopeful, named the sculpture after Charlie Quartpot, a traditional rainmaker.

This statue gave Mark none of the flak he later got with the McDouall Stuart statue, which had him asking himself if it was worth it. 'Like you'd think it would be so easy—it's just a bit of artwork,' he said at the time of the protest in Alice Springs. 'If I took [the gun] away, what would he be holding, a bunch of flowers?'

The Aileron roadhouse owner, Greg Dick, when asked why he commissioned the statue, is quoted as saying 'Well, a fucking minke whale or a dolphin would look out of place out here.'

Central Mount Stuart is next on the map, also an asterisk on the poster hanging on my bedroom wall back in Adelaide. McDouall Stuart recorded on his fourth expedition in April 1860: 'I find from my observations of the sun, that I am now camped in the centre of Australia.' He'd pinpointed this spot, 111° 00' 30", using a sextant to calculate the centre of the continent, equidistant from east to west and north to south. His calculation was a little bit out, but I'm still impressed.

While spinifex and thick scrub plagued McDouall Stuart as he made his way to the centre, tearing saddlebags and clothes to pieces, I've got it easy. I'm experiencing the spinifex grasslands, mulga and mallee scrub from the comfort of Betsie's cabin.

The map on the passenger seat beside me is folded to highlight the centre and all I have to do is take a sharp turn left into the lay-by off the highway, where there's a memorial to McDouall Stuart's achievement. I would have missed it if it hadn't been for a busload of tourists, binoculars at the ready and facing the highway, not the memorial, which did strike me as odd.

Reaching the centre was the dream that McDouall Stuart had kept on the back burner for sixteen years after his stint on Captain

Charles Sturt's famous inland expedition in 1844-45. Captain Sturt had hoped to be the first European to plant a flag at Australia's geographical centre, but that expedition never got close. Another hope was to find the inland sea that so many people were convinced existed in the middle of the Australian continent. Sturt was so confident that he took along a wooden boat and a couple of sailors, but his mainsail remained forever furled. His health broken, and probably heartbroken too, he returned to England never to come back.

'I completely forgot aboot taking a flag tae mark the moment,' McDouall Stuart admits to me.

'A bit remiss of you.'

He ignores me. *'It wis Esther Knowles, the manager's wife back at Moolooloo who saved the day. She made one for me. Widnae take any money. Telt me tae name a hill after her. Which I did.'*

McDouall Stuart was a bit put out that the nearest mountain to plant the flag on was a few miles away from his calculated centre. He and Kekwick trekked to the foot of the mountain and 'after a deal of labour, slips, and knocks', they arrived at the top. He was to write in his journal:

> '[We] built a large cone of stones, in the centre of which I placed a pole with the British flag nailed to it. Near the top of the cone, I placed a small bottle, in which there is a slip of paper, with our signatures.
>
> We gave three hearty cheers for the flag, the emblem of civil and religious liberty, and may it be a sign to the natives that the dawn of liberty, civilisation, and Christianity is about to break upon them.'

Captain Sturt, who hired McDouall Stuart as mapmaker and

surveyor for his expedition and then elevated him to second-in-command, was to say later: 'I am not at all surprised at Stuart's success for I know him to be a plucky little fellow—cool, persevering and intelligent, as well as an excellent bushman. He has fairly passed, or, I should say, surpassed me, and may justly claim the laurels.'

Ten years or so later, another Scottish surveyor, John Ross, marking out the route for the Overland Telegraph Line, spotted what was left of the flagpole and recovered the note left by McDouall Stuart. No sign of the flag itself. This communication line was the start of the full effects of colonisation coming to the centre of Australia, forty years after South Australia was proclaimed, and eighty years after the First Fleet arrived to establish the convict settlement at Botany Bay in January 1788.

'Ye ken, I named that mountain Central Mount Sturt, tae honour my auld boss. It wis changed tae Stuart later. Wisnae my idea'.

I've no intention of trying to climb Central Mount Stuart. It's enough for me to look at it from a distance and to read the poster in the lay-by. Here I meet a pair of modern-day explorers—a young couple who've cycled from their home in Hungary to Australia. They are using the notice board to hang up their water filtering tubes. McDouall Stuart could have done with that gadget. 'Examined a large creek; can find no surface water, but got some by scratching in the sand,' he wrote in his journal before climbing up the poorly placed mountain. I'm the luckiest of all. I've got clean, cold water in my car fridge.

The Hungarian couple's gear, compact and light, is impressive, especially a nifty solar-powered mat with a USB slot for charging up their other gadgets. I thought I was flash with a solar-powered torch. While they and I have been comparing the past and the present in solar power, a vision of solar power's future potential comes silently gliding towards us. In the shape of a pencil-thin vehicle, with a head bobbing around under a glass bubble, it's the leader in the annual World Solar Challenge race from Darwin to Adelaide. Which explains the line-up of onlookers.

I donate my cold water and some fruit to the plucky couple as I'll be in Tennant Creek before long. They head south and I head north. McDouall Stuart had headed west. A bad idea.

He and his two companions were aiming for what was later named the Tanami Desert. It's one of the most isolated and most arid areas on earth. McDouall Stuart was nearly killed when a wallaby spooked his horse, which then dragged him along the ground, kicked him at a fearful rate, and struck him on the shoulder. That injury troubled him for the rest of the expedition.

Almost every entry for those three weeks in that desert spelled disaster and doom, except for one uplifting moment when he praised Kekwick and Head, writing: 'They have exerted themselves to the utmost, and everything has been done with the greatest alacrity and cheerfulness. Although they have only had two hours of sleep during the last two nights, there has not been a single word of dissatisfaction from either of them, which is highly gratifying to me. It is, indeed, a great pleasure to have men that will do their work without grumbling.'

By the time they trekked back to Central Mount Stuart, the horses were in a bad way, Kekwick was suffering from bad hands, Ben Head had taken ill a few days earlier and McDouall Stuart wrote in his journal: 'For the past three weeks I have been suffering dreadful pains in the muscles, caused by the scurvy; but the last two nights they have been most excruciating, so much so, that I almost wished that death would come and release me from my torture.'

Wikipedia calls the Tanami 'a final frontier' not fully explored by white Australians until well into the twentieth century. Today's travellers are advised to carry plenty of drinking water as all dams and bores along the route are classified as undrinkable and the longest stretch without a fuel stop is nudging 600 kilometres.

The next dot on the map is Barrow Creek, once a repeater station for the Overland Telegraph Line, seventy kilometres further up the highway. The population of around ten includes another larger-than-life roadhouse owner, Les Pilton, who calls his establishment the Pilton Hilton. 'Bullshit and beer' is the offering on the billboard.

Near here is Skull Creek, the site of a massacre in 1874 where 50 to 60 Aboriginal men, women and children were killed. So much for the civilised and Christian behaviour that McDouall Stuart had wished upon the locals when he planted that flag back at Central Mount Stuart less than fifteen years earlier.

The story told in the government information sheet for tourists at Barrow Creek has a photograph of a headstone marking the grave of two workers on the Overland Telegraph Line. The story is of the men at the repeater station being attacked (no reason stated), followed by a series of 'reprisal expeditions' headed by the police (no mention of how many Aboriginal people were murdered). To find information about Skull Creek, you have to go online to check the 'Colonial Frontier Massacres map'. It relates many massacre stories that have a similar pattern: white men abduct and rape Aboriginal women; Aboriginal men retaliate and spear men they consider to blame; a band of white men gather, often involving the police, and what happens next is called 'teaching a lesson'.

Barrow Creek today is possibly best known in Australia and overseas as the closest settlement to where, in 2001, English traveller Peter Falconio was lured from the campervan he and his girlfriend

Joanne Lees were driving north on the Stuart Highway. Peter's body has never been found. Les Pilton gave refuge to Joanne in the days after the murder.

The day I drive into Barrow Creek, there's no talk of colonial-era frontier violence or kidnapping and murder at the turn of this century. All attention is on the clutch of solar car challengers bedding down for the night.

I continue travelling north, while McDouall Stuart's journals reveal he veered off his planned route as he had spotted a mountain, named it Mount Strzelecki after fellow explorer Count Strzelecki, and decided he had to climb it.

'What is it with you and bumps on the landscape?' I ask as I move up through the gears and settle into a steady pace. That side trip was a chore just to get to the foot of Mount Strzelecki, the men struggling through very thick mulga scrub and pushing the horses past clumps of spinifex. I've had a few run-ins with that jaggy little fucker and it's never pleasant.

'Maybe you should have been satisfied with just naming it from afar?'

'I ken. I suffered mightily for doin' it, I huvtae say.'

'And then, when you reached the top, you helped to build the cone of stones.'

'It wis terrible killin work.'

'Couldn't you have let Kekwick and Head do that?'

'My party wis too small to leave it to them.'

I end my day behind the Devils Marbles Hotel, hoping to find a bit of shade. The temperature around here can go above 40 degrees Celsius some days, and today feels like one of those days.

I visit the Karlu Karlu / Devils Marbles Conservation Reserve nearby where gigantic balls of granite are precariously balanced on top of one another and strewn across the landscape. This is a significant sacred site for the Warumungu, Kaytetye, Alyawarre and Warlpiri people and was officially given back to the traditional owners in 2008. Karlu Karlu translates to 'round boulders'.

There's no getting away from those solar car drivers. When I see the sumptuous dinner for the Japanese team being prepared in a fully-equipped mobile kitchen next to my mozzie dome, I decide my one-pot noodles will stay in the packet. I'm dining on a schnitzel and a beer in the pub. I know how to treat myself.

Perched on a stool, I chat to yet another backpacker working their way around Australia, this time Scandinavian, who tells me she's swapped her job as a nanny on a cattle station to pull pints. It was a lonely existence on the station, she says.

A character sidles up to us. 'You do know we're only a few kilometres from where Peter Falconio was kidnapped and killed.'

The barmaid looks startled.

'Fifteen years ago,' I reassure her. If it's not going through Australia's impressive list of deadly beasties, tales of backpackers coming a cropper from creepy criminals in remote locations is a favourite conversation starter with unsuspecting overseas tourists.

Territorians enjoy their larrikin image, the reputation for not taking themselves too seriously and engaging in bullshitting at the bar. Massive termite mounds along the Stuart Highway are dressed in all kinds of clobber and signs designed for a laugh. I'm not sure if this tourism advertisement is still doing the rounds: CU in the NT. I rest my case.

TWENTY-SIX

DON'T TALK ABOUT THE WAR

The Stuart Highway slices through the middle of Tennant Creek township, scattering buildings on either side.

A government tourism website calls this 'a friendly hamlet'. Don't know about you, but the word 'hamlet' for me conjures up an English country garden sort of scene. What I see is a dinkum dusty outback town; shops strung along either side of the main road, their wide verandas suffering heat stress and offering questionable protection from the blistering sun.

'Welfare town' is the label it's given by national and online media, which tend to favour the shock factor, along with terms like 'high crime rates', 'social dysfunction', 'substance abuse' and 'endemic poverty'.

Only 3,000 people live here, but it's by far the largest town you'll pass through on the 1,000-kilometre-plus haul from Alice Springs up to Katherine. As the first European to walk over this land back in 1860, McDouall Stuart's journals suggest he once again took the opportunity for a bit of naming around the place, though I do question whether he controlled all the choosing. I can't picture him having time on his expeditions to mull: 'Now, which upstanding citizen will I give a little posterity to here?'

Tennant Creek, the waterway, (called Jurnkkurakurr by the local Warumungu people), he named after one of the expedition financiers from Port Lincoln. The Bonney, he named after a South Australian politician, and yet another creek, the name escapes me, is allotted to a random bank manager in Adelaide. McDouall Stuart considered the Bonney to be 'the finest creek I've come across since Chambers Creek'. I can sort of see why, even though the channel, wide and long, is crackling dry when I cross over it.

I pay for three days at the caravan park and set up camp.

'Three days? You're up tae somethin', aren't ye?'

I don't know what to say. Seemed a good idea at the time. Truth be known, I'm not sure what I'm up to. Driving along Stuart Highway, I'd nothing better to do than mull over my inability to find an Aboriginal person to talk to in Alice Springs about McDouall Stuart. I was thinking I could give it another go in Tennant Creek, where the population is two-thirds Aboriginal. Attack Creek, the site of the confrontation between McDouall Stuart's band of three and local Warumungu warriors, and where his fourth expedition came to a halt, is only seventy kilometres up the highway. He described the warriors at the time as formidable foes: 'bold and daring... wily and determined'.

McDouall Stuart was right when he predicted gold mining in the region, but he didn't foresee what colonisation would look like for the wily, determined Warumungu warriors at Attack Creek.

By the 1890s, once the Overland Telegraph Line was built and pastoralists were spreading north, Warumungu people were living in camps around the Tennant Creek repeater station, some receiving rations, while others worked there.

In 1892 a reserve had been set aside for the Warumungu but that was revoked in the 1930s to clear the way for gold prospecting. Overlooking the town is the Battery Hill Mining Centre and a monument in town of a miner panning for gold commemorates Australia's last gold rush. Men came from all parts of Australia hoping to make their fortunes, the stockman's track was upgraded to partially gravelled to cope and the population of Tennant Creek township grew.

By the 1960s, the Warumungu had been entirely removed from their native land.

As I settle in my swag, I go over all the reasons why this is not such a good idea. 'Don't fuck it up,' says the voice in my head. Then up pops a scene from Berlin, the year after the Wall came down. My sister and I were there, sitting on a bench looking over to a tenement-looking building when we saw an elderly man limping towards the entrance with its security access panel of buttons. 'Don't talk about the war,' my sister warned me. Which was going to be hard as I wanted a peek inside the building where we had lived after the Second World War. German families had been turfed out to make way for the likes of us, families of soldiers in the British occupying forces.

In my very basic schoolgirl German, I asked the old man if he could let me into the stairwell, and I got the gist that he was visiting his son, so sorry he couldn't. And then, I swear, he started chatting about why he was having trouble walking. He'd been shot in the war, he said, slapping his thigh. I gave my sister one of those 'what were you worrying about' looks. Decades later, I've convinced myself he may have invited us in for a cuppa if I'd persisted.

I've paid for three nights camping, I remind myself, so the inner me must be sure of herself. Time to gird my loins.

My quest to find an Aboriginal person to talk to begins with the local newspapers in the local library, a safe option to avoid any chance of offending anyone or being snubbed. Old copies of The Tennant and District Times newspapers are full of articles about McDouall Stuart's fourth expedition, for the 150th anniversary in 2010. There are whole columns, week after week, excerpts from McDouall Stuart's journal, summaries from historians and reports of various celebrations for me to photocopy.

McDouall Stuart enthusiasts in the town travelled 300-plus kilometres south to Central Mount Stuart for a special event to mark him being the first to reach that point. There was one front-page story of an event in Tennant Creek by local Warumungu people, who chose to commemorate the expedition with a horse trail ride, with local senior men returning to the saddle to honour their ancestors who had worked as stockmen before them. The article says the aim was to show respect to those old men who were up each day before dawn and worked until after sunset, seven days a week, for payment in rations of tea, sugar and flour.

I head out of the library with a sheaf of photocopies and toss them on the passenger seat to read later, along with brochures that focus on cattle kings and mining. Mention of Aboriginal culture focuses on describing a way of life before colonisation, involving Dreaming stories and bush tucker and then jumps to information about Aboriginal art, dancing and music that visitors can experience now. As I put Bad-Ass Betsie into reverse, I'm thinking about our settler-colonial history that lies between those two pictures, and the phrase, 'the Great Australian Silence' that describes this gaping hole.

A jolt knocks the pile of papers onto the floor and the unwashed dishes are sent sliding and clattering across the back seat. The crumpling sound I heard would explain the orange plastic I'm now staring at on the ground. The rear light cover is broken and the bumper is looking the worse for wear too. I've backed into a parked tray-top truck that doesn't display one scratch. At least, not one caused by me.

'Sorry Betsie,' I say as I rub her dented bumper. On the way up the highway, I'd sailed past evidence of other people's misfortune: long, swerving skid marks and strips of scattered rubber; abandoned cars, some still with wheels, others red-rusted and pock-marked. I avoided bounding kangaroos and lumbering bullocks. Not a scratch did Bad-Ass Betsie have when I rolled into town, a stationary Council truck breaking my accident-free run.

Book borrowers have come outside to give advice so I'm meeting the locals right enough, but this wasn't quite what I had in mind. The police could stop me and defect the car, predicts a young lad, who says that's what happened to his big brother. I nod and his big brother stuffs the bulb back in its socket, shaking his head. 'Not likely,' is all he says.

I slink into the Council office, more to escape attention than anything. The lass behind the desk who has been watching the action fills in my details on the accident report form. 'Didn't like the look of our animal rescue truck, eh?' She suggests I head to the police station to report the incident.

'I didn't like the look of the Council's rescue truck.' The line was so amusing I thought it worth repeating at the police station, but the whippersnapper behind the counter isn't amused. Surely this boy is masquerading as a police officer. He keeps his eye on me as he reaches under the counter.

'Don't touch. Just blow.' He stares in disbelief at the blinking zeros on the breathalyser. I really, really want to make him stand in the corner until he learns to treat his elders with respect.

TWENTY-SEVEN

WARUMUNGU WARRIORS

'We weren't happy about Stuart taking the water—but later we lost our land.' I'm reading the first inscription for a series of dioramas in the Nyinkka Nyunyu Art and Cultural Centre. It's next to a scene of McDouall Stuart passing through the region.

Other dioramas, a feature of the centre since it was opened in 2003, depict the saga of dispossession: pastoralism, the Overland Telegraph Line, mining, the missions and the birth of the town.

'The ideas for each of our dioramas have come from our community,' I read. 'They tell different parts of our story, in the way that we like it to be told. There are some sad stories and some good stories here.'

I chat with a few people at the centre about my quest and spend some time absorbed in gathering as much info as I can. Nyinkka Nyunyu is a mixture of a museum to share special artefacts and information with tourists, a place where the young ones in the town can be educated about their culture, and a place for local and visiting artists to paint and exhibit. 'Each of our museum collections shares unique parts of who we are and where we have come from,' the website says.

I walk past shelves of bush tucker, edible seeds, fruit and medicinal products. A gargle to soothe tooth pain is made from boiling the bark of the thorny bush plum. I photograph a display of tools and weapons, returned home after being taken away or given as gifts hundreds of years ago, many stored in museums around the world. Colourful posters represent Warumungu 'punttu', which is often translated into English as 'skin name'. Each Warumungu person has one, maybe two punttu, and there are eight women's names starting with 'N' while the eight men's names start with 'J'. You can't change your name, just like you can't change your skin.

'You should talk to Ross Williams,' I'm advised as I prepare to leave the centre. 'He was in a TV documentary about the Overland Telegraph Line and there's a scene about Attack Creek. Played the role of one of his warrior ancestors.' McDouall Stuart didn't find any gold in these parts when he passed through, but I feel like I just have.

'Just go to one of the pubs on the main street. Ask for him at the bar and someone'll go and get him.' Before I can chicken out, I drive down the main street. Tennant Creek is within the vast semi-arid savannah plains of the Barkly Tableland, said to have some of the best cattle grazing country in Australia. The stockman era is on my mind, explaining why it's the wide-brimmed hats and checked shirts that I'm drawn to in the main street. No one's walking very far or very fast in this heat.

I park in the middle of a row of well-loved utes and troopies outside one of the pubs. Bad-Ass Betsie, smeared with red dirt and in need of a good wash, looks at home next to them. She also looks like she means business with that mine light on top, and I'm hoping I look as capable as she does.

It's dim inside the pub, the light coming from the far side where the room opens to the beer garden beyond. I stride to the bar where a young woman is wiping beer glasses. If she says he's not here, I'll just leave and that will be it. She's determined to be helpful. 'Hey Joe,' she calls out to a group over at the pool table. 'Know of a Ross Williams?' The click of balls stops for a second before starting up again.

A second lass joins us. The pair are from Belfast, backpacking around Australia. They have been working here for two weeks, having a great time, and are already on first-name terms with most of the punters. But they've not heard the name Ross Williams. The balls start clicking again and we have that conversation when you hear a familiar accent from your part of the world: 'Where are you from? How long have you been out here?' 'Really? That long? Bloody great, isn't it?' 'You've kept your accent. Good on you.' 'We're having a ball.' 'Me too.' 'Aussies don't know how good they've got it.'

'It's Jakamarra you're after.' I turn round. This must be Joe. He does that sign that Aboriginal folk I've worked with have perfected—a tiny pout you're not sure you saw at all. Joe's looking towards the far side of the bar that leads out to the beer garden.

The man looking over at us wasn't there a few seconds ago.

'My name is Ross Williams. I'm a Warumungu man. My skin name's Jakamarra. My totem is the white cockatoo.' This information shows connection with the land. Not for the first time I feel I'm missing out somehow, not having that closeness, not in Scotland, the country I voluntarily left, and not here.

We're propped on stools at a small round table in the beer garden out the back and I've placed my tape recorder between us. Ross nods to the tape recorder and I turn it on. He's smiling: 'I've got Scottish ancestors too.'

It's a few years now since Ross worked with historians in the making of the ABC TV documentary *A Wire through the Heart*, about the construction of the Overland Telegraph Line.

'For me and my nephews to have the chance to do that, to imagine what these old warriors were going through, it was an experience and a half. And a privilege too,' he says.

'To live through what our old people might have lived through. It makes you proud. I could just imagine what it would have been like for them warriors back then, seeing my first white man, and the horses, you know, it would have been frightening, thinking "What am I going to do?"

Plus, they were letting the horses drink our water, the water that we own. That would have driven those old warriors to attack. They would have said, "How come they think they have a right to do that?" I'd have done the same thing in that situation.

'And for McDouall Stuart and his men too. I would say they were panicking seeing those spears flying, and they started firing back I suppose.'

'Yes, the whole thing would have been very hard. Shock on both sides. Ignorance on both sides.'

A group of men standing off to the side are waiting to talk to Ross who is a respected elder in town and a traditional owner. 'Always things to discuss about Warumungu land, Warumungu country,' says Ross and he excuses himself.

While he's gone, I think of all the reasons given for why the attack took place: It followed a confrontation over dwindling water supplies; fear; a misunderstanding across the language barrier. One of the men from the group walks over and introduces himself as Ross's brother. I can't resist asking what he thinks happened at Attack Creek.

'I wasn't there,' says Kenny. 'I only know the stories. I don't know the facts.' And he's so right. I've heard so many conflicting stories, tying myself in knots trying to work things out, weighing one version against another. Ross returns and the conversation turns to various projects in the town, which I take as a signal that the topic of Attack Creek is finished with.

'We've got the local language centre, to keep the language alive, how to speak the correct Warumungu language, the strong language,' Ross says. 'Only about four or five people speak it strongly now. I understand it, and I speak some of it. A lot of the elders produced Warumungu dictionaries. It was a big job. My older sister was involved in that.'

Ross, born at Phillip Creek pastoral station and growing up at another, called Banka Banka, tells me his station days were 'pretty good', with memories of his dad building mud huts, his mum working as a maid, and the children having chores to do after school.

The story of another local, Harry Bennett, written in the local newspaper, tells of a different path to Banka Banka. After a childhood in the bush travelling with his family, setting up camp, hunting and following food sources and natural water, Harry suddenly found himself taken from his family, ending up in The Bungalow in Alice Springs, becoming a member of the Stolen Generation. This was in the mid-1930s. As an adult, he had many different jobs, finally moving to Banka Banka in the 1960s, where he got involved in supporting the land rights movement. 'That was the end of working on the stations,' he says in the article.

I sense the dioramas back at the cultural centre are like Aboriginal paintings that depict an artist's country, with their multiple layers of hidden meaning and history behind each scene. A code that's hard to crack.

I don't want to overstay my welcome and bid Ross farewell. But I can't help myself. A comment in Alice Springs niggles, another perspective to commemorating McDouall Stuart as the first explorer to map the land.

'Do you think some people might blame McDouall Stuart for all the changes that came through years later?'

'I don't think so,' Ross says. 'I think it was going to happen sooner or later. Inevitable. Can't be stopped. And it eventually happened.

'Not so long ago, until the 1960s, we weren't allowed to come into Tennant Creek,' he adds.

'Now we have Native Title over more than half the town. A bit ironic, isn't it?'

In 2007, Tennant Creek became the first town in Australia to have a Native Title determination reached without prolonged and costly court hearings. This Native Title agreement allows for large areas of the town to be developed. In return, the Patta Warumungu people have freehold title to land outside the township, including the Devils Pebbles significant site in the north.

Back in 1978, the process wasn't so agreeable when traditional owners claimed under the *Land Rights Act* for land outside the town boundaries.

In an attempt to make large tracts of land unclaimable, the Northern Territory government's tactic was to try and extend the boundaries of Tennant Creek to make the town fifteen times bigger. The legal battle lasted fifteen years, one of the most protracted land rights claims ever in the Northern Territory, before the traditional owners got full ownership of a strip of country stretching a hundred kilometres north and south of the town.

The *Land Rights Act*, only applicable in the Northern Territory and stronger than Native Title, legally recognises the Aboriginal system of land ownership since before colonisation and allows for ownership of land not already owned or leased to pastoralists or other interests.

Traditional owners in the Northern Territory to date have won back around 400,000 square kilometres of the Northern Territory.

TWENTY-EIGHT

FORCED TO RETURN

McDouall Stuart wondered if he was now in territory already crossed by explorers. He writes in his journal about a couple of encounters they had with locals, who didn't seem shocked to see them. Going by his maps, it could have been Gregory travelling west to east across the Top End some years earlier.

The first encounter involved a couple of strapping young men who met the three expeditioners and handed them a few possums, some small birds and a parrot.

'That encounter didnae end well,' says McDouall Stuart.

'They wanted tae steal awthing they could lay their fingers on. I caught one concealin the rasp for shoeing the horses. He put it under the netting roond his waist. I hudtae take it from him by force.

'They wur determined tae have the canteens as well. Cheeky blighters. I had a helluva job getting them back. They wanted tae pry into everything until I lost all patience and ordered them off.'

'We had a braw supper with the possums that night, mind you, better

than we'd had for many a day.'

He hadn't cottoned on that perhaps the Aboriginal pair were expecting something in return for the food, and it would have been good manners anyway.

No wonder he spoke about a slap-up meal. Their diet of flour, week after week, was pretty basic and his journal shows he'd been worried about Kekwick and Head, who were complaining of weakness from the lack of decent food.

Then came the second encounter.

'One of the lads returned wi an old man and whit a surprise. He gave me a Freemason sign,' says McDouall Stuart.

'Are you sure it was a Freemason sign?' I ask. 'Pretty indiscreet, whoever taught it to the old man, wouldn't you say? Weren't the Freemason signs supposed to be top secret, passed onto fellow members only?'

He ignores me. *'I just stared at him, and returned the sign and you ken what happened next?'* I shake my head. *'He patted me on the shoulder and stroked doon my beard. Then he walked off, friendly as can be.'*

Two days later, with no more encounters, McDouall Stuart records that he was thinking the natives had left them in peace, 'as natives in general do, in consequence of having seen us pass in the morning.' That was not to be.

One of the weekly articles in Tennant Creek's local newspaper, written by author Stuart Traynor to commemorate McDouall Stuart reaching the centre of Australia, refers to what happened next. 'They continued riding back along the creek but then were confronted by about thirty warriors who shouted at them, threatened them with boomerangs and set fire to the grass around them. Stuart tried to appear friendly but noticed even more warriors emerging from behind the bushes. After a couple of volleys of boomerangs, one of which hit Stuart's horse, the three men opened fire. Stuart does not say in his journal if any Aboriginal people were wounded or killed. They had no option but to retreat southwards with the warriors following them some way, yelling and setting fire to the grass. It was 11 o'clock that night before they got to Hayward Creek and felt it was safe to stop and camp.'

After considering the matter over the whole night, McDouall Stuart's journal entry the next morning is as follows:

'I have most reluctantly come to the determination of abandoning the attempt... to make the Gulf of Carpentaria... my party being far too small to cope with such wily, determined natives as those we have just encountered... With such as these for enemies in our rear, and most probably far worse in advance, it would be destruction to all my party for me to attempt to go on.'

The three of them had scurvy, supplies were very short, the horses were in poor condition, and the country was drying out. They were facing a 2,400-kilometre slog back to Adelaide.

The simple inscription on a plaque at Attack Creek to mark the centenary of McDouall Stuart reaching this spot reads:

HOSTILE NATIVES AND ILLNESS
FORCED THE PARTY TO RETURN.

'You were at death's door, you were on your way back down, and yet I can see from your journal that you still hankered to have another go.'

'*I ken. If only it would rain, that's what I thocht at the time. I kept hopin the weather wid turn. Be kinder. Ma hopes kept gettin' raised. Then dashed.*'

'You were running out of food too.'

'*I had a thocht tae kill one of the horses and dry his flesh. That wid've been enough to get us back.*'

McDouall Stuart finally accepted he was heading home and retreated to Bonney Creek. The place he so cheerily named on the way up was now the site where he admits: 'A sad, sad disappointment; all our most sanguine hopes are again gone, for, during the night, the clouds broke up and have all vanished... I shall rest the horses til Monday, and then, ill and dispirited, commence my homeward journey.'

They were a sorry sight as they trudged home. The horses had sores on their backs that wouldn't heal and McDouall Stuart was worried about Kekwick and Head. 'My men have now lost all their former energy and activity, and move about as if they were a hundred years old,' he wrote.

His journal entries were weighted with disappointment. 'Had I but a stronger party, and six months' rations, I think I should [have been] able to accomplish something before my return,' he lamented. 'I have done my best, and can do no more.'

They picked native cucumbers by the bucketload to help with the scurvy, boiling them up with some sugar. Tasted a bit like gooseberries, he reckoned.

'*All the way back doon, I was workin oot how to get back up again, mair men, mair horses the next time, that's for sure.*'

And more horseshoes too, I was hoping.

Finally, they were back over the border and at Moolooloo. McDouall Stuart wrote a letter to Chambers in Adelaide: 'I am sorry. I have been unable to make the north-west coast [Victoria River]—the difficulties have been more than I was able to overcome.'

Adelaide newspapers and politicians were having none of that, calling McDouall Stuart's achievement both daring and heroic under the headline:

EXPLORATION OF THE INTERIOR
TRIUMPHANT JOURNEY OF MR STUART

It was hardly surprising that the newspapers were hyping up Stuart's achievements. Or that Parliament, previously lukewarm towards McDouall Stuart and his backers, was now more than keen to get on board, putting their hands in their pockets for the first time to get him out there again as soon as possible. Burke and Wills, leading the Victorian Exploring Expedition, had left Melbourne a month before, also aiming to reach the northern coast.

Adelaide was in danger of losing its status as the first port of call for news, with letters and newspapers arriving by boat from the motherland. Whoever succeeded in becoming the first colony to reach the north coast to win the contract for the Overland Telegraph Line would take over that prestigious honour.

The Adelaide press and politicians tied themselves in knots to convince everyone that not reaching the north coast was a moot point, splitting hairs. What a convoluted argument they spouted to claim that, by reaching Latitude 18 degrees, McDouall Stuart and his two companions had near as dammit joined the dots with explorers travelling across the top. This meant they could proclaim to the world that:

But McDouall Stuart didn't feel like a hero. Sure, he'd got to the centre, 'unlocked its secrets', but he was smarting at not making it to the top. Near enough was not good enough for him and 'almost' was a word to beat yourself up over.

Flush with its success in the goldfields, Victoria had opened its bulging purse for Robert O'Hara Burke, giving him £12,000 to get fully kitted out. The expedition consisted of eighteen men, twenty-seven camels and twenty-three horses pulling drays loaded with twenty-one tons of provisions. This was the first major expedition to use camels as a means of transport. The load included four dozen fishing lines, nine hundred and fifty-four sets of camel shoes, ten dozen looking glasses, eighty pairs of shoes, several cases of surgical equipment, and an extensive armoury. Also considered essential were thirty cabbage-tree hats, a dining table and a Chinese gong to summon diners, a library of books, eight demi-johns of lime juice to ward off scurvy, four gallons of brandy for the men and sixty gallons of rum for the camels. The rum to rub on the camels' footpads didn't make the final cut. The brandy for the men did.

The South Australian government gave McDouall Stuart £2,500 to spend and he buckled down to plan the trip. I reckon he'd have gone back out there again, money or no money.

I was pleased to see he packed away 150 sets of horseshoes and plenty of nails.

'Aye, and you'll note I took along a shoesmith.'

The Government was offering a reward of £2,000 to anyone who could get to the top on a route through the centre, rather than promising it directly to McDouall Stuart.

'I bet that was galling,' I say.

He shrugged. *'Through the centre, they said. I was the only one doing that.'*

This was to be McDouall Stuart's best-provisioned expedition so far, ten men and forty-four horses, but still modest. He felt no need to upgrade his navigation equipment; he was perfectly happy with his compass, sextant, extendable telescope and astronomical almanacs.

Frugal as ever and sticking with his expedition style of travelling light, McDouall Stuart was hard pushed to spend half the money allotted.

1860 THE GREAT AUSTRALIAN EXPLORATION RACE
The challenge was on to secure the route for the proposed
Overland Telegraph Line. Victoria was banking on Robert
O'Hara Burke, he of the bushy beard, while South
Australia was betting on McDouall Stuart whose facial
hair could get a bit on the straggly side.[1]

1. '1860 THE GREAT AUSTRALIAN EXPLORATION RACE.', Melbourne Punch (Vic. : 1855 - 1900), 8 November, p. 4., viewed 25 Sep 2023, http://nla.gov.au/nla.news-article174525940

PART 5

THE

TROPICS

from Adelaide
to Attack Creek
to Longreach Waterhole

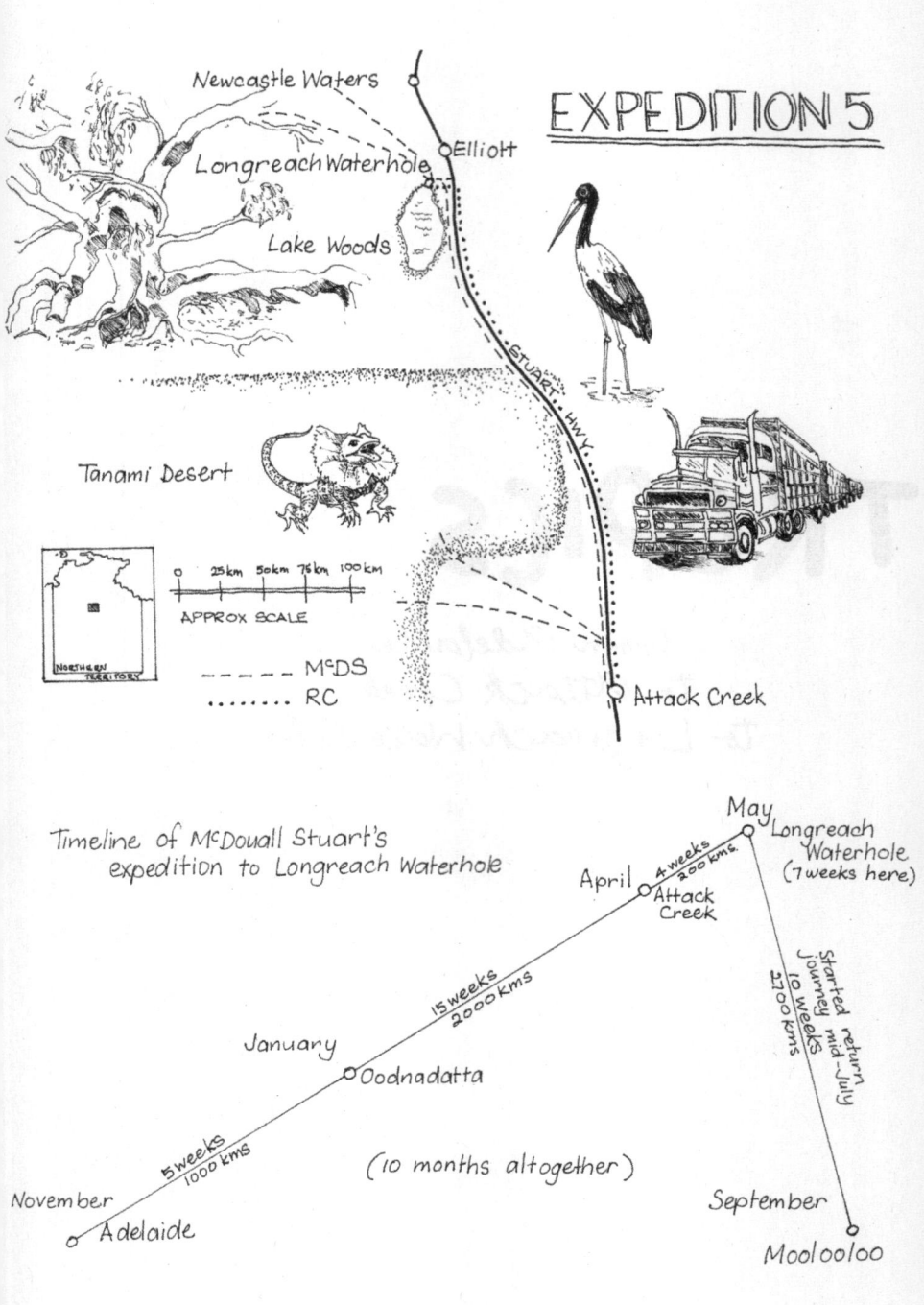

EXPEDITION 5

Newcastle Waters

Elliott

Longreach Waterhole

Lake Woods

STUART HWY.

Tanami Desert

0 25km 50km 76km 100km
APPROX SCALE

NORTHERN TERRITORY

— — — McDS
.... RC

Attack Creek

Timeline of McDouall Stuart's
expedition to Longreach Waterhole

May
Longreach
Waterhole
(7 weeks here)

April
4 weeks
200 kms.
Attack
Creek

15 weeks
2000 kms

Started return
journey mid-July
10 weeks
1700 kms

January
Oodnadatta

(10 months altogether)

5 weeks
1000 kms

November
Adelaide

September
Moolooloo

TWENTY-NINE

A BIG BLUE BLOTCH

'There wis never ony competition wi' Burke and Wills, ye ken.'
'That's not what the newspapers were saying.'
'Aye, yer right there, hen.'
'And the government... it was keen to win the race too.'
'Aye, I ken. Disnae make it true though.'

Before they'd even left the settled regions of South Australia, horse problems befell McDouall Stuart's expedition. The town horses he was given were soft and nearly useless, getting 'knocked up' far too easily, he complained, and he swapped some over with new ones as soon as he could—they turned out to be so fresh, they kept bolting.

Accepting that some of the horses were not coping, he decided to dump provisions around where Oodnadatta is today, calculating that food for thirty weeks would suffice to get them to the top and back. After all, he and Kekwick knew the stretch up to Attack Creek well, so it wouldn't take them long. This was a calculation he would come to regret.

Experienced horses, out with McDouall Stuart on previous expeditions, were suffering on this trip, too. One morning, Bennett, one of his best packhorses, one he could depend on for a hard push, was found dead, either from bad water or from eating something poisonous. 'This is a very great loss to me,' McDouall Stuart wrote in his journal, but he knew sentimentality wasn't going to get them through this calamity. 'As we were very much in want of hobble straps, I sent Mr. Kekwick with three others to take Bennett's skin and shoes off,' he wrote. Around the same time, Polly, his favourite horse, had a miscarriage. She was given one day to recover.

Everyone benefitted from a rest, though, when another horse wandered off and spent the night up to his body in water. He was trembling all over when he was found and rescued, scarcely able to walk. 'I should gain nothing by starting today,' McDouall Stuart noted.

As they approached Attack Creek again, the air was heavy and the soft ground was not helping McDouall Stuart's mood. One day, three horses went missing because of the scarcity of feed, wasting precious time, and some horses were becoming so weak they were unable to carry a decent load.

His objective had been to have the horses as fit and healthy as possible at this point of the journey, ready for the unexplored country ahead. Instead, he was writing in his journal: 'The sweat was running in streams from the horses and, when we halted for a few minutes, they were puffing and blowing as though they had just come in from a race.'

Fifteen weeks it took to get back to Attack Creek. By now, they were supposed to have reached the top. Burke and Wills were somewhere to the east, goodness knows where.

One good thing. They slinked past Attack Creek, this time without any encounters.

McDouall Stuart was determined to join up with the path taken by Gregory, to indeed join up the dots of explorers who previously trekked across the Top End, but he was too far east for his liking. 'No hope of success on this course,' was his verdict two days out from Attack Creek, a conclusion he was to say, over and over, each time he changed direction.

Here was a man who loved the desert: these tropics were something else. If it wasn't a dense scrub of dwarf lancewood as tough as a whalebone that was before them, it was rotten ground with large deep holes and cracks hidden under long thick grass. Booby traps for the horses.

It would be madness to take the horses another day over such a country, he said a week later at the end of 'the hardest and most fatiguing day's work since starting from Chambers Creek'. He considered himself lucky to escape injury that day when his horse stumbled into a hole and nearly rolled on top of him. The absence of all birds was not a good sign.

Another week followed with more changes of direction. McDouall Stuart was astounded at how heavy the ground had become, the horses at times sinking in the black alluvial soil over their knees. 'In the short distance we penetrated, it has torn our hands, faces, clothes and, what is of more consequence, our saddle bags, all to pieces,' he wrote.

McDouall Stuart was banking on his knowledge of birds and their flight paths to lead him onwards to water and made a note when he saw some pigeons. The next day, they saw turkeys after sunset, upwards of fifty passing over in twos and threes. The following day it was a large flock of pelicans 'which leads me to think that there may be a lake in the vicinity,' he wrote.

My map shows a big blue blotch, 180 kilometres up the Stuart Highway from Attack Creek, at the end of a wiggly track heading west. That blue blotch doesn't fool me, though. Throughout this journey, rivers and creeks, optimistically marked in blue on the map, have turned out to be but a hint, a distant recollection of water once upon a time. The creeks I'm driving past are dry and sandy indentations.

It took me a couple of hours driving to reach the turn-off. It took McDouall Stuart four weeks to reach what he called 'a splendid reach of water' before naming it Glandfield Lagoon, a name later changed to Newcastle Waters. Today's map lists it as Lake Woods, and the top corner is Longreach Waterhole. Yeh. Right. That was him writing in 1861. I don't care how many watery names it has, I'm not expecting much. I haven't seen a decent stretch of water since the swimming pool in Alice Springs.

Daylight is fading as I get to the end of that track and roll Betsie up to the water's edge. It's like a massive bolt of crushed silver taffeta rippling across the landscape. I was wrong. Longreach Waterhole is indeed splendid. It is part of the Lake Woods wetland conservation area for scores of bird species that I should know some names for, but don't. It's a refuge in the dry season, supporting over 100,000 waterbirds at times, and is an important stop-over for migratory waders.

McDouall Stuart's descriptions when he got to these waters were almost poetic: of pelicans in the air, ducks and white cranes gliding on the water, sacred ibis stepping along the shore, and mounds of mussel and periwinkle shells on the banks. His men checked the depth of water and found it was six feet deep, ten steps from the shore, dropping to seventeen feet in the middle.

They caught fish resembling whiting and he recorded: 'I had one cooked for tea; the skin was as tough as a piece of leather, but the inside was really good, as fine a fish as I have ever eaten.'

If you ask me, that sounds like hard work. I can't even be bothered setting up the gas burner, a decision that I'll be thankful for very soon.

I busy myself taking photos of the pelicans as they slide across the huge expanse of water, small groups in turn stretching and reaching for the sky, as synchronized as a dance routine, their flapping wings creating a mini whir.

Dusk descends and the rosy-crimson stripes of cloud overhead are mirrored in the waters below. The sky turns deep red and a silhouette on the shoreline could be an ibis. It all looks so serene, the water slapping the shore as I settle in my swag for the night and I wait for stardust to sprinkle across the skies. A moment of bliss.

Background rumbling builds up then dies away—like the lullaby of distant traffic, but there's no traffic for miles. Now it's the murmur of flying foxes flitting through the canopy of trees overhead. A sudden gasp of wind swirls a moat of leaves around me. This is exciting.

Pitter, pitter, pitter. Fat raindrops are plopping through the netting. I yank the groundsheet from beneath the mattress, drape it overhead and peg it to the poles of my flimsy dome. A decent make-do tarpaulin if I say so myself.

Those bats know better than me. They're now screeching to each other, careening through the treetops in search of shelter. The wind howls, the lightning flickers close by and the rain, at first in cupfuls, is coming in bucket-loads.

That tarpaulin is not decent and will not make do. Ping, ping, ping go the puny clothes pegs and the tarp somersaults off into the trees, leaving me clinging to the poles as they twitch and contort and threaten to rip what's been my home. I'm marooned.

The poles whip free from my grasp and the netting of the mozzie dome, frantic in newfound freedom, whips across my face. Something bumps against me. I hope it's nothing important. Or scary.

I'm surrounded by thousands of leeches, stretching up and up, becoming as thin as needles, straining for my blood. Dread and fear start to nibble the edges of my imagination which has the lake waters rising tsunami-style to sweep me downstream. Where I am is just a corner of Lake Woods. The whole thing covers 350 square kilometres, sometimes double in the wet season and treble in full flood.

The storm ends as quickly as it began. The downpour becomes a drizzle, a flimsy curtain of fairy lights falling through the trees and then it stops altogether. The lap-lap-lap from the water's edge is the only other sound as the wind runs out of puff.

Two choices. I could sit in the driver's seat of the car, upright and sodden, or I could clamber into the back, perhaps garrotting myself on wayward string or skewering my palm on an upturned knife. I take neither. In the light of the half moon, I surface and resurrect the poles and spend the night clinging to the edge of the sodden mattress with a warm, wet bottom. I'm whimpering like a baby needing its nappy changed. I smell the dampness all around. Leeches, go your hardest. I don't give a fuck any more.

I had what you'd call a fitful night.

THIRTY

GETTING NOWHERE

The watery-white early morning sun is already heating up as I drape all things sodden over the car doors, the bonnet, the bumper bar and the tailgate: a steamy lakeside laundry. I retrieve the tarp which had wrapped itself around a distant tree and pick up the scattered clothes pegs. Leave only footprints, remember.

I walk along the shoreline past gnarly old trees that look how I feel. A line of elegant white cows saunter along the opposite bank, pause and look over. It's me, watching them, watching me. Time for a morning brew but I don't wait for the water to boil: tepid tea will do. I'll be leaving as soon as everything's dry.

I pull out my notebook to jot down details of last night and imagine McDouall Stuart back in 1861, sitting outside his tent, maybe close to where I am now, writing by the light of the moon, or a campfire, or maybe the flicker of a tallow candle, night after night, recording important information.

No matter how splendid McDouall Stuart thought this spot was, he had no intention of delaying here a minute longer than necessary on his fifth expedition. The latitude at this spot is 17 degrees, 36 minutes, 40 seconds. He aimed to get to 18 degrees and a wee bit further west, where explorer Gregory had his camp a few years earlier. He'd been told to head that way if he could, as it was still the preferred route for the proposed Overland Telegraph Line.

'It wis very disheartening work. I'm sure ye could tell from what I wrote.'

I sure could. Time and again the lament was 'I shall try... once more.' And there was the oft-repeated lament: 'I am now running short of horseshoes.' I pour him a cuppa.

'I didnae get to Mr Gregory's camp. There wisnae the least appearance of rising ground, or a change in the country—nothin but the same dismal dreary forest throughout. The horses widnae face it.

'Oh, how I wished for something to carry water. I should then be able to dae it.'

McDouall Stuart's two leatherbound notebooks from his fifth expedition were small enough to slip into his trouser pocket. Before this trip, I'd taken to requesting them in advance on research visits to the Royal Geographical Society library back in Adelaide. Not to read them. I couldn't. The writing is so tiny. All I wanted was to run my fingers over the covers and marvel at the perfectly straight lines of copperplate script running from edge to edge of each page as he made use of every skerrick of space. I remember thinking he could have a second career writing on grains of rice. I had to delve into printed copies to look for the gems that would reveal the man hidden within the minutiae of wind direction, land conditions and horseshoe issues. I take my hat off to the resilience of researchers. I nodded off every time.

As a distraction, I'd seek out McDouall Stuart memorabilia: a piece of cloth in the muted colours of the Stuart of Bute tartan; a large framed print of him with sleeves rolled up, planting the Union Jack in the centre of the continent. At other times I'd slide the faded pink ribbons from dog-eared folders and spread their contents over the embossed-leather table at the back of the room; handwritten letters and sheaves of flimsy typewritten pages clipped together. Snippets gave me glimpses of the man, the explorer.

I got used to McDouall Stuart looking over my shoulder. The miniature statue in the corner of the room, eerily grey and almost translucent, makes him look quite the man about town. His beard, clipped and curled into ringlets, has the hallmark of a visit to the barber shop and his cloak is jauntily draped over one shoulder. I've seen photos of him. No way did he pose for that statue.

One day I picked up the library's mighty magnifying glass and tried to make out some dates and some words in the first leather-bound notebook. I'd read a printed version but this made me feel so much closer to the action. The writing got smaller and smaller as he recorded his forays, day after day, getting nowhere. I jumped to the end of the notebook. The date is July 1. 'Not a drop of water or a watercourse have we seen... It is hopeless to proceed further... All hope of gaining the gulf without wells is now gone.'

The first entry in the second notebook had him setting off on another foray, taking ten horses, three men and a month's provisions, leaving the rest of them at the camp. That was his strategy every time. He wasn't about to give up. 'I may succeed. I am very unwilling to return without trying all that is in my power.'

One week later, however, it's:

'We are all nearly naked, the scrub has been so severe on our clothes; one can scarcely tell the original colour of a single garment, everything is so patched. Our boots are also gone.

'It certainly is a great disappointment to me not to be able to get through, but I believe I have left nothing untried that has been in my power. I have tried to make the gulf and river, both before rain fell, and immediately after it had fallen; but the results were the same.'

One word at the end of this second journal, dated July 11, is in capital letters and underlined:

'UNSUCCESSFUL'

'I want to make it clear, I named Burke's Creek after ma brither explorer, just afore I turned back for hame. Along the road frae here.' McDouall Stuart waves his hand in a general easterly direction.

'I wis thinking at the time 'Where can he be?' He prods my papers. 'Put that in yer jotter.'

'Our readers,' the Adelaide newspapers falsely reported, 'will, we are sure, be glad to hear, on what may be considered good authority, that Mr Stuart and his party have returned in safety after having accomplished their trip across the continent.'

'Got a bit ahead o' themselves.' McDouall Stuart holds out his mug for a refill. The press and the politicians had done it again, spruiking McDouall Stuart's 'brilliant success' and applauding him for being so humble in victory. Getting it wrong.

'Mr Stuart is, in fact, the Napoleon of explorers,' the Register newspaper intoned, keen to focus on McDouall Stuart's accomplishment by 'simple and unostentatious means.' The article suggested this constituted a new era in exploration.

'Embarrassing. I planned tae tak thirty weeks in total tae get to the north coast and back. We were away for forty weeks. And didn't get much further than before.'

A correction of sorts was printed a few days later. 'We stated that Mr Stuart had crossed the continent, but this does not appear to have been precisely the case.' In the parliament, it was stated that McDouall Stuart's inability to reach the top was 'trifling indeed, compared to what he has accomplished,'

When he arrived back in Adelaide in late September, Governor MacDonnell suggested a public shindig to present him with a medal from the Royal Geographical Society of London for getting to the centre on his previous expedition. No thanks, said McDouall Stuart. I wasn't surprised. Not one for being in the spotlight was our Mr Stuart, even though this made him the second person, after Dr David Livingstone, to receive both a watch and a medal from the Society. It was reported that Governor MacDonnell accepted that 'the less ceremony observed the more acceptable it would be to Mr Stuart' and so it was that McDouall Stuart agreed to accept the medal behind closed doors.

Meanwhile, the Burke and Wills story was unravelling with reports that some of the men on that expedition had perished and the whereabouts of those two was unknown. McDouall Stuart offered to go and look for them, but search parties were already out looking.

McDouall Stuart was required to head back out straight away. That meant travelling in summer, but the haste suited him fine. He never was one to linger long in the city between his forays.

PART 6
EXPEDITION

Expedition Six
The Great Northern Exploring Expedition
January 1862 – December 1862

SIX

From Adelaide to Point Stuart

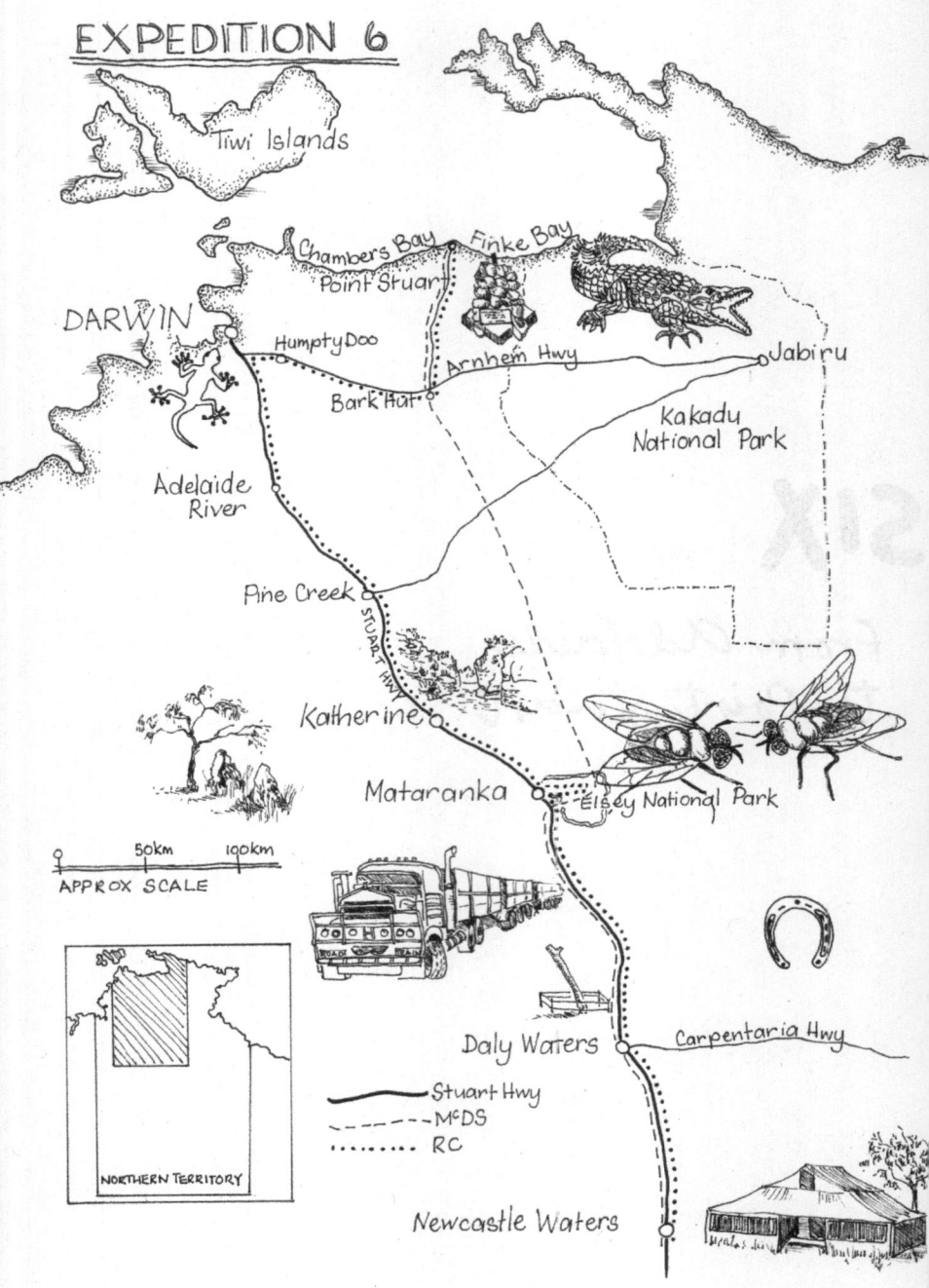

EXPEDITION 6

Tiwi Islands

Chambers Bay
Point Stuart
Finke Bay

DARWIN

Humpty Doo

Arnhem Hwy

Jabiru

Bark Hut

Kakadu
National Park

Adelaide
River

Pine Creek

STUART HWY

Katherine

Mataranka

Elsey National Park

APPROX SCALE

50km   100km

NORTHERN TERRITORY

Daly Waters

Carpentaria Hwy

Stuart Hwy
McDS
RC

Newcastle Waters

THIRTY-ONE

WHAT A PARTY THAT WAS

On Friday 25 October 1861, the number of toasts at the farewell luncheon for McDouall Stuart's Great Northern Exploring Expedition was impressive. They sure knew back then how to wring out every opportunity to clink glasses at formal occasions. The locals lined up along the boundary of Montefiore House in North Adelaide, the home of James Chambers, to watch the proceedings.

'I give you this day nine months for I know how many difficulties you will encounter,' said state politician John Bentham Neales. 'Stick to your leader, boys, and when you return, we will give a cheer that will make your ears tingle.' These rousing words were from Police Inspector George Hamilton, no doubt followed by another round of glass-clinking.

McDouall Stuart had taken one month to hire his team for this trip, with Kekwick and Woodforde, both on the previous expedition, already back at Chambers Creek drying the bullock meat and packing it away.

Police inspector Hamilton had spruiked his invention of canvas water bags to McDouall Stuart. They'd give horses the camel advantage, he said. McDouall Stuart was impressed. Chambers' daughter Elizabeth had sewn JMDS in the middle of a Union Jack flag for McDouall Stuart to plant in the soil when he reached the northern shores. He liked that idea too.

Sons of up-and-coming Adelaide establishment families were among the recruits for the sixth expedition. As explorers were a bit like today's popstars, these young men were no doubt counting on the girls swooning on their return.

Stephen King Junior, whose father was one of the colony's most successful early pioneers, was one of the first to be signed up. He was given the job of breaking in the new horses. James Frew's father was a pioneer merchant and estate agent and Heath Nash was the son of Dr James Nash, the first colonial surgeon of South Australia. Nash knew how to make damper so he was the designated cook.

Pat Auld was a late addition to the group. He thought the expedition sounded like a jolly good idea when he heard about it from his old school chum King on the night before the farewell luncheon. The cocky lad presented himself to Chambers and McDouall Stuart. They told him to report for duty at 5 am the next day and off he went to town to buy himself boots, breeches and blankets for the heady world of exploring.

Francis Thring, at 24, was the oldest among this lot and the only one at the luncheon who had any idea what was in store as he had been on McDouall Stuart's previous expedition. He was to be in charge of the horses and saddles while John McGorrerey was the team's shoesmith. Loitering on the fringe of the group was John Billiatt, King's cousin, 19 years old and the youngest of them all. Just out from England and smitten with King's sister, he was tagging along in the hope of eventually being accepted into the pack, no doubt to impress his new love.

'Whit a party that wis.' McDouall Stuart is enjoying reminiscing and I'm resigned to sit here at Longreach for a while longer. The view over to that waterhole could be a lot worse. The billy goes back on and out comes my jotter where I've pasted quotes and my notes, about the farewell event.

> **Well-oiled after all the toasts, the lads found it difficult to keep upright when it was time to mount the horses and head off. The crowd assembled on the boundary of James Chambers' home in North Adelaide to wave them off witnessed quite a spectacle. Riders were falling off their steeds and packhorses bolted, scattering bags, horseshoes and swags in all directions.**

A bit of a boy's own adventure, what-ho.

That scribble in the margin wasn't meant for McDouall Stuart's beady eye, but too late.

'Oh. And what's this adventure you're havin' yersel', hen? An auld lassie's adventure?

'Less of the 'old', thank you.' I top up his mug. 'You know Auld dined out for years on that undignified departure? He's got an interesting turn of phrase, don't you think?'

'Oh aye, he was a wag was that one.'

I show him snippets I'd taken from Auld's 'Recollections of McDouall Stuart', printed in newspapers and repeated by Auld every chance he got for years afterwards. Here's a beauty.

'One horse broke into a garden and Frew and I rode in to drive him out, when a very excitable lady appeared on the veranda and, with broom in hand, expressed a strong desire to brush the dust from our jackets.

'Had I been a poor orphan and had she expressed a desire to adopt me, I would have preferred going alone to being under her protection.'

Auld was no orphan, of course. And not poor. His father was the owner of what was to become Auldana winery, one of the largest in the colony.

'Those young lads could knock it back, I can tell ye. By the time we left I could see there'd be a few sair heids the next day.'

They all set off along the North Road towards the Heart and Hand hostelry.

'"One for the road, lads," is what I said. Seemed a good idea at the time. I'm looking roond and there's a bit of a kerfuffle. Horses gettin restless... men gettin restless... crowd gettin restless. "Och. For god's sake man, give me that rein," said I. And I tried to unfankle the rope. Well. Anither guid idea and I don't think.'

I've got a slip of paper with the newspaper clipping of that event – and another snippy comment from me in the margin.

> 'The horse reared and struck Mr Stuart on the temple with its fore foot, knocking him down and rendering him insensible. The brute then sprang forward and placed one of his hind feet on Mr. Stuart's right hand, and, rearing again, dislocated two joints of his first finger, tearing the flesh and nail from it, and injuring the bone to such an extent that amputation of the finger was at first thought unavoidable.'

That would've sobered the rest of them up.

While the rest of them continued north, McDouall Stuart returned to Adelaide for treatment and it was five weeks before he could set off to join them.

'I had a sair hand, but ma injuries wirnae as bad as the newspapers made oot.'

I'd read, however, that there were fears he wouldn't be able to lead the expedition at all after that accident. He admitted later that the injury gave him grief the whole journey. The accident had maimed him for life.

On the eve of McDouall Stuart finally setting off, his arm in a sling, news came that Burke and Wills had perished. In fact, they were already dead before McDouall Stuart had turned back for home on his previous expedition.

THIRTY-TWO

A MIDLIFE CRISIS

McDouall Stuart wasn't discouraged by his accident. Neither was he daunted by the five times he previously had to turn back, nor fazed by the Burke and Wills tragedy.

'That's some midlife crisis you were having.'

'Midlife crisis, eh? What aboot you, hen? What are you having...a later-on-in-life crisis?'

I ignore him.

McDouall Stuart had his instructions. His priority was to find a route for the proposed overland telegraph line. The eyes of the world were on him and he must not be foiled again, the newspapers said. On no account was the goal to be abandoned. No pressure then.

'You do know your obsession was killing you.'

'I wisnae gettin' any younger. Me and Waterhouse were 46, the oldest in the team.' The government had wanted naturalist Frederick George Waterhouse, an Oxford-educated entomologist and the first curator of the South Australian Museum, to go on the previous

expedition. But McDouall Stuart and Chambers had refused. This time the government got its way and McDouall Stuart was told to give Waterhouse 'every facility' to carry out his collections and examinations of the country.

'People were saying you didn't like Waterhouse.'

'I didnae want a botanist, naturalist, whatever his role was, bein' on the expedition, that's all. Holdin' us up, gettin' mair money than the rest of the men, and me no' able tae gie him ony chores. I hoped he'd decide no tae come along.'

'I know the feeling.'

'Is that right, noo? Someone has to keep an eye on ye. A smile's lurking under all that facial hair. 'D'you remember gettin to Burra? Two days it took you to drive there. How far is it frae Adelaide? I was wonderin' if you were ever going to get over the border, never mind up tae here'.

'Very funny.'

'That guy in Burra winked at you. Dae ye mind him? I reckon he wis flirtin.

'Flirting? What would you know about flirting.' I could wind him up with the story I'd heard about him thinking he'd been jilted back in Scotland before he emigrated. But we were starting to get along not too badly, so I kept my mouth shut. I couldn't resist a wee dig, though.

'That guy was talking about you and your drinking, if you remember.'

'Here we go. Aye. I like a drink, who didnae in Adelaide? But for some reason ma drinking didnae please some folk. Ah well. Ah don't think I harmed onybody else in the doin'. Just masel... I bet you're goin' to tell the story noo aboot Moolooloo, eh?'

'I wasn't actually.'

The Moolooloo story I *was* going to tell was about Auld. He'd been left there for a month to wait for Waterhouse and McDouall Stuart to arrive. He wrote later that the cook said there were more jam tins opened while he was there than in the previous four years and he left that station with the smell of jam in his nose. I remember the smell of lanolin in my nose after *my* time at Moolooloo, spent in the woolshed.

'But now that you mention it... is that the story about you getting, how shall I put it, tired and emotional? So much so, you were unable to give the talk you were supposed to give at the local Institute. I believe Waterhouse had to take over, is that right?'

'Ma very own PR person, ye promised. Goin tae bring me the attention you thocht I deserved, ye said. Some attention this is. For the record, hen, I never touched a dram once I was beyond the settled regions... by the way, I do hope yer no intending tae make me the excuse fer yer ane drinkin.'

Ouch.

McDouall Stuart, Waterhouse and Auld joined the rest of the team at Chambers Creek, which is what McDouall Stuart insisted on calling this land in his journals.

I imagine him sitting around the campfire listening as the lads recalled the adventures they had on the way up, often about the horses. One got spooked by a whirlwind, another bolted at the sight of a flock of emus.

'It was Hogmanay when you got there, last chance for a knees-up, but no celebrations. Call yourself a Scot?'

'We wurnae there for a party.'

He probably wasn't in the mood, right enough. For one thing, he wasn't happy that the men were using flour bags to pack the jerked beef. And so New Year's Eve and the following week were spent drying the last of the bullock meat and making new bags.

'I mind one of the lads told a story aboot an auld biddy at Arkaba. She wis overheard sayin she wis sure such a lot of beardless boys would ne'er be able to cross the continent. "We shall never see them back," she said. Well, I wisnae goin' tae let that happen.'

McDouall Stuart was proud that, so far, he had not lost one man to accident, illness or violence in any of his expeditions. And so, on the morning of 8 January 1862, as they were about to set off, McDouall Stuart, for the first time on any of his expeditions, handed out a list of rules.

Thring's copy of the regulations, filling three pages, are in State Records of South Australia (SRSA GRG35/636).

'Quite a list you made there.'

'Aye. I wisnae takin' any chances and the rules made it clear I wis havin' nae nonsense.'

One rule was about not firing on the natives without orders unless in self-defence. I wondered if the Attack Creek incident had changed the tone of his expeditions. Before Attack Creek he'd said something about there being no need for big guns, now I read that he'd been given gifts of revolvers and rifles and they were 'armed and ready'. There seemed to be a mood change in the newspapers and parliament, too. There was the occasional lonesome voice of dissent but the more usual comments were that there was a need to be prepared, that the natives had no right to the land and that nothing should stop colonisation and 'progress'. A number of rules referred to the horses, to make sure they were well cared for, and, of course, he had one specific rule that 'No swearing or improper language shall be allowed'.

By the end of January, the expeditioners had reached Peake, near Oodnadatta, still the most northerly station in South Australia. McDouall Stuart kept his word about not putting up with insolence or insubordination. Jeffries the saddler had left the party near Chambers Creek and Woodforde was heading home. Billiatt got his wish. He was now officially part of the group.

This was the last opportunity for the lads to write home before heading into the desert and their letters read like modern-day gap-year blog posts, full of enthusiasm and wonder at their new surroundings. No more stations, not even an outstation nor a shepherd's hut, 'away from the smoke of a chimney, all dray tracks and white men and entirely on our own', King wrote.

Auld's letter described their routine, each day starting with packing the horses, each pair of men in charge of eighteen. McDouall Stuart would light his pipe, Auld wrote, take a bearing with his prismatic compass and off they'd go, McDouall Stuart in front,

Kekwick at the rear.

'Wait til they read this one,' you can almost hear King saying as he wrote about their arrival at Louden Springs. They had been impressed when, in the middle of the night, McDouall Stuart stopped to announce that a spring was nearby so, after hours of riding in hot oppressive weather, the horses and men made a rush for the water. 'When daylight came, we were very much disgusted at the stuff we had been drinking,' King wrote. They'd been drinking, not from the spring, but from a nearby soak of muddy water, full of horsehair, emu feathers and a dead dingo.

It's the end of February. Rations of flour, meat, sugar and tea had already been halved to lighten their load, the greatcoats had been chucked away, and now it was another two hundredweight of sugar unloaded. The weather, hot as ever, got a mention in McDouall Stuart's every journal entry and it was clear that encounters with locals were escalating. 'I am afraid they are going to be a very great annoyance this time,' McDouall Stuart wrote.

The horses were having a rough time, too. One that was poorly had died along the way, another good horse drowned and many had worms. Five more gave in as they crossed the South Australian border.

It's March and McDouall Stuart and his men were crossing the MacDonnell Ranges. 'The horses knocking up is very vexatious and a serious hindrance to my progress,' he wrote. 'If all the horses had been equal to my former ones, I should have been at the centre a fortnight ago.'

He was always concerned when he saw native smokes and it was at Mount Hay that he noted at length his defensive reaction to some incidents. One was when they faced a group of seven natives yelling and shaking their spears while separating themselves in a half-circle. His journal entry is that he had his party prepare to receive an attack 'but when they saw us stationary, they approached no nearer. I ordered some of the party to fire close to them, to show them we could injure them at a long distance if they continued to annoy us, but they did not seem at all frightened at the report of the rifles nor the whizzing of the balls near to them, since they remained in a threatening attitude'.

In the same area, McDouall Stuart records that Thring had been forced to use his revolver in self-defence when he came across three natives hiding behind a bush about to attack him with spears. And a few days later, when they saw another group placing their spears in their woomeras preparing to attack, he ordered Auld to fire at the rock they were standing on, which 'sent them off at full speed'.

Peter Latz in Alice Springs had suggested the rising tensions could be partly attributed to word spreading fast into the centre of the continent about what had happened to Aboriginal people elsewhere: stories of violent clashes with pastoralists, miners and settlers pushing them to the fringes of their land, annihilating their way of life. They were right to be wary, Peter Latz said.

McDouall Stuart made it clear in his journal that he too was wary, and wrote he was prepared to make an example of the 'natives' if they threatened his men again because 'it is evident their designs are hostile'.

Weeks later than planned, they arrived at Central Mount Stuart once more, though McDouall Stuart didn't mention the occasion in his journals, so intent was he on keeping going. They also slipped past Attack Creek without incident.

Then it was back here to Longreach.

'Three months tae get back here and I wis determined no' tae hang aroond for long. One week, for the men and horses tae get their strength back.

'The last time when I tried tae move frae here, the horses wur withoot water for days at a time. Once it wis for mair than a hundred hours. I had high hopes for thon waterbags. They wur to save the day.'

'And how well did they work for you?'

'Could've gone better.'

The one-week plan for Longreach turned into more than two months with eleven forays out and back again, some lasting a week or so, some just a couple of days. McDouall Stuart moved base four times, upholding his routine of never advancing until water was found. He led several treks west into the Sturt Plains, his map showing a spread of fingers reaching out to a river that proved out of reach.

'I dinnae ken what made me try again for the Victoria River. Some days we wur makin little mair than a mile an hour. The further I went, the denser it became.

'Maybe it wis because I could see that three or four wells would do the dab, overcome the want of water for the overland telegraph line. And the forest, well, the trees would be fine for the poles. I believed it could be penetrated, a matter of cuttin a line through and burnin it.'

Boredom set in for those left back in camp while McDouall Stuart and a few hand-picked men went searching for a way forward, so they were happy to entertain a group of natives who visited often.

On one occasion, Kekwick brought out some tape and calico. 'We had great fun decorating the youngsters, who seemed delighted,'

Auld reminisced later. 'One young fellow positively blushed and placed his hand before his mouth when a looking glass was held up in front of him.'

The Bible was the only book they had, so the men contented themselves with telling each other stories, and the topic of food was a constant. Fantasies about dinner parties and going into restaurants, astonishing everyone by the quantities they ate, and still not being satisfied.

On one occasion, they were so hungry, they didn't think twice about eating a litter of dingo puppies. According to Auld, these innocents were slaughtered, boiled and pronounced delicious. Another time, a dog was shot, skinned, cut up, put in the pot and eaten, the day after some damper had gone missing. 'Some of the men's palates were so sensitive that they solemnly declared they detected the lost damper as they ate,' Auld wrote.

By the time McDouall Stuart decided to head off, water or no water ahead, the men didn't need to be told twice.

It's time for me to head off too. There's a town I want to see called Newcastle Waters. It came long after McDouall Stuart was here. And it withered and died before I arrived in Australia.

THIRTY-THREE

DROVING AND DRIVING

Newcastle Waters township, destined for greatness if you listened to the dreamers, but dying when droving came to a halt. It's now described as deserted, obsolete, abandoned, uninhabited.

A green wooden cross nailed to the corrugated iron wall is the only clue that the first building I walk over to was once the town church. Gaping windows staggered along the wall are minus the louvres that once tried to control the climate for Sunday service. Today the yawning rectangles invite eye-piercing sunlight to throw geometric patterns over the cracked and crumbling floor that's more in keeping with a car mechanic's workshop than a place of worship. No stained glass, no pews, no altar. As I take photos of the arty light show, I swear I can hear the faint tinkling of children singing.

In the late 1800s, this was a dusty spot on the Australian map, a meeting point for overlanders droving mobs of cattle from pastoral stations to distant markets on the east coast. A feisty Irishwoman called Catherine McCarthy saw potential and opened for business in a ramshackle tin shed, offering hot meals 'on an hour's notice' to hungry drovers passing through. And so the township was born. A police station was built in 1914, followed by various structures built with mud bricks and sheets of corrugated iron along the one and only road.

There was talk at one time of Newcastle Waters becoming a major aviation hub for the north. Another idea in some heads was for the proposed railway from Darwin to come to town. Yet another, that the Stuart Highway, slicing through the centre of the country, just might pass through. A short-lived attempt to divide the Territory into two had Newcastle Waters earmarked as the capital of the northern section. Or at least a major administrative centre.

But the township's reason to breathe was already slipping away by the time this church was built by the Aboriginal Inland Mission in 1964. The days of droving were all but over, as shifting beef on the hoof was replaced by road trains transporting bellowing cattle along the Stuart Highway (which was always going to bypass Newcastle Waters). The Junction Hotel served its last beer in 1960, and closed its doors for good in 1976.

A row of buildings remain, some bowed and battered into submission, others well-preserved, serving as museum pieces to show visitors the harsh and hardy life of early pioneers.

Sounds of laughter *are* coming from the building in the next paddock. I'm not imagining it. Before I have a chance to climb the steps, the door at the top swings open and two little people are peering out at me. A woman steps from behind.

'You're not the first to think it's a ghost town.' Rosemary Raymond, who became Principal of Marlinja (Newcastle Waters) School in 2005, sits with me at a bench outside the classrooms, the only cluster of buildings still in use in Newcastle Waters. I see now that this building has had a lick of paint.

Rosemary was born in the nearby Marlinja Homelands Aboriginal community and was a student herself at the school. She points to a building where, in the 60s, she and her classmates used to line up for showers before class, and then she walks me over to a small building, now a storeroom, where there's a large iron ring bolted to the cement floor.

'When my dad was a student here in the 30s, this room was used as a jail cell. They used to tie prisoners up in here.'

The school teaches a dozen or so students from the community, sometimes with the addition of children of station workers at the huge Newcastle Waters pastoral property, once owned by media mogul the late Kerry Packer.

'The school is very important for the community,' says Rosemary. 'We try to keep the culture alive, teach language. We have our ceremonies in Marlinja, teach them corroboree dancing but... kids are starting to lose their culture... too much influence in the mainstream. They'd rather watch TV than go dancing.' I nod. We agree it's a common problem across all cultures to keep children interested in their heritage.

'The kids read about Captain Cook, that he discovered Australia, and they think white folk were here first.' Rosemary laughs at the look on my face. 'I straighten them up, don't you worry. I tell them we are the first people of Australia.'

Rosemary has to get back to work and I walk over to the Drovers Memorial Park, a well-tended grassy area across the road from the school, dominated by a larger-than-life bronze statue of a rather dapper drover.

The engraving on a plaque beneath the drover reads:

As the stock are slowly stringing,
Clancy rides behind them singing,
for the drover's life has pleasures
that the townfolk never know.

Clancy of the overflow
—A.B. Paterson

The drover statue was unveiled in 1988, the year Australia commemorated the landing of the First Fleet in Botany Bay.

No wonder the local kids are confused.

Over at some information boards are details of McDouall Stuart coming through this area, including extracts from his journals of encounters with the locals. Jingili are the people he would have met, Rosemary had suggested when I mentioned him.

'*We saw them, often, for sure.*' McDouall Stuart pipes up. '*Everywhere along the track and farther afield. Mostly it wis clues that they'd been here, or they wur peekin at us from a distance. I huvtae say, some encounters started off badly and improved, some went the ither way.*'

He points to a quote on the board about Aboriginal people coming to see them, sitting down and having a good stare.

'*I mind that ane. That wis on the sixth expedition and they stayed a long time withoot showin' ony inclination tae go. And this ane,*' he points to another quote on the billboard.

> **This morning we were visited by seven natives—tall, powerfully built fellows... shaking their boomerangs. We made friendly signs to them to come nearer, and they gradually approached.**

'That wis definitely on the fifth. I remember it well. Here's anither. Not sure when this wis.'

> Once more I have returned... we were again visited by our black friends... One of the natives was an amusing little fellow... He imitated every movement we made, and burlesqued them to a high degree, causing great laughter to his companions and to us.

The Muranji Track, embedded in Northern Territory folklore as part of Australia's droving history, is also mentioned on the information board.

'Made a name for itsel', I see. Suicide Track.'

That's one name listed here. Ghost Road of the Drovers is another, and Death Track. They leave nothing to the imagination. The 230-kilometre stretch connecting Newcastle Waters to Victoria River was one of the most treacherous stock routes in Australia, with no watering points and a fierce reputation.

'Treacherous, did you say? Tell me somethin' I dinnae ken.'

The sinking of bores, sometime after the First World War, made a difference, but it was still tough. Drovers faced dense jungles of bulwaddy and lancewood scrub, and told tales of cattle getting spooked and stampeding after hearing the echoes of their hooves as they walked over the top of underground limestone caverns.

Wave Hill pastoral station, at the far end of the Muranji Track, is the site of the historic Gurindji walk-off by more than two hundred Aboriginal stockmen and their families. It made headlines across Australia as a protest about wages and conditions, but it was always a pivotal moment in the Aboriginal land rights timeline, leading to land rights legislation for First Nations people in the Northern Territory. The photograph of the prime minister in 1975, Gough Whitlam, pouring soil into the hand of Aboriginal elder Vincent Lingiari, helped to put land rights on the national political agenda.

There's no mention of land rights in Drover's Park.

On the other side of the drover's statue is an installation of a giant open book unveiled in 2012. One page commemorates the 150th anniversary of McDouall Stuart's explorations in this region. The facing page says this land, given by the Dreamtime spirit ancestors, is where the shared boundaries of the Mudbura from the west and the Jingili from the east intertwine. This is well-watered country, an oasis in a dry land, and related clans in more arid areas beyond also have spiritual ties to this area. The installation mentions past gatherings 'from time to time'.

There's no commemoration marking the major gathering in 2012 to celebrate the Mudbura and Jingili winning Native Title over 30,000 square kilometres of their traditional lands.

McDouall Stuart's journal entries about meeting Aboriginal people were used in that lengthy legal process lasting twelve years. His entries helped the Mudbura and Jingili people prove they have managed to keep a continuous connection to their lands for tens of thousands of years and Newcastle Waters Station was the first working pastoral station in the Northern Territory to have co-existing Native Title rights recognised.

There's no mention of that in Drover's Park.

THIRTY-FOUR

A BANQUET OF PARROTS AND TEA

Back on the Stuart Highway, I hear the rumble of road trains, but I'm learning to fight the impulse to slow down. A monster truck pounds past, the whiff of sweaty cattle trailing behind. I count three trailers, six decks in all, and it's lurching. Hard to credit a bigger, longer beast is allowed on the roads—a four-trailer seven-deck combo.

No highway for McDouall Stuart. As he pushed north from Newcastle Waters, he'd pull out his binoculars on the constant lookout for water. Not a puddle escaped posterity as he named each discovery in honour of one of his men: McGorrerey Ponds, Auld's Chain of Ponds, King's Pond, Frew's Waterhole, Nash Spring. He wrote about pushing through thick woods, lancewood scrub and tangled forests of bulwaddy, and described ground so hard that the hooves of the horses scarcely left an impression.

It would be ten years after McDouall Stuart came through before the track he trod became a supply road for the construction workers on the Overland Telegraph Line. And ten years more for the sometimes-boggy, sometimes-bulldusty track to become the north-south stock route for the first pastoral stations in the Top End.

The pastoral industry has moved from droving to road trains; from boundary riders to helicopter pilots and aerial mustering; from isolation to two-way radio and satellite communications. Hundreds of road trains each year shift a million or so head of cattle in, out and around the Northern Territory, heading for an abattoir, a port or a pastoral station and a lot of the action is on the Stuart Highway, now part of a 36,000-kilometre national network of sealed roads.

'I wis travelling a mile an hour, back in the day,' groaned McDouall Stuart as a second road train overtakes us. I'm on the lookout for the turn-off to Daly Waters which is called a town despite its tiny population.

A slogging twenty-two days it took for McDouall Stuart to get here. A couple of hours it took me—and that includes multiple stops to photograph towering red termite mounds, each one rising higher than the one before, up to twice my height. Some stand among patches of blackened trees that resemble scorched statues, others are wrapped around fallen tree limbs criss-crossed in piles like abandoned funeral pyres. I can't tell if I'm seeing the aftermath of an out-of-control bushfire, or deliberate burning.

First Nations people traditionally used firestick management to domesticate plants, influence animal movement and prevent uncontrolled wildfires. Lighting small controlled fires was part of everyday life. Today, looking after country includes the traditional, age-old knowledge of fire with cutting-edge techniques and equipment.

As he approached Daly Waters, McDouall Stuart was happy with what he saw ahead: 'thickly wooded with tall mulga and lancewood scrub to the west—but to the east open gum forests—splendidly grassed'. His way was clear. He and his men camped around here and, according to his journal, they celebrated Queen Victoria's birthday with a banquet of parrots and tea. This would be a brief stop. Or so he thought.

My brief stop is at the Daly Waters Pub, which has the longest continuous liquor licence in the Northern Territory, beginning in 1893. Today's version of the Daly Waters Pub opened in 1930 and still has the hitching post outside for the drovers passing through.

Two busloads of tourists beat me through the door as I admire the exterior—blazing-red bougainvillea flowing over the front veranda. But I still get to the bar first. The tourists were side-swiped by the over-the-top interior décor, particularly taken they were with the bunting with bumps. It's said the bras hanging down from the rafters were from women losing a no-holds-barred beer-drinking game.

Plastered on the walls, posts and ceiling—any surface within reach—are business cards and banknotes from all around the world, photos and sage sayings and mementoes left by visitors, including an Irish hurling stick.

Down the road a bit, Australia's first international airport is easy to overlook. The runway was first used in 1930 by Amy Johnson, the first woman to fly solo from England to Australia. Domestic and international Qantas flights once stopped here to refuel. In the 1930s it took three days for planes to fly over the Australian continent. During the Second World War, the hanger was a staging post for Australian and American air force personnel.

A short stagger-distance from the Daly Waters Pub is Stuart's Stump leaning over like a punter who's had one too many. Bare and branchless, this still-life vision does have some minimalist appeal. On 23 May 1862, McDouall Stuart—or one of his men—carved the letter 'S' into this tree trunk, says a plaque, and the proof is a photo taken in the 1940s of said snaky 'S'. I can't find any carving. But it's hot and I don't look that hard.

This is my favourite McDouall Stuart memorial, though it uses probably the most unflattering photo of the man that I've seen yet. This memorial was erected thanks to those wartime air force officers who made sure this relic from his explorations was surrounded by a low-flung fence, validating its importance.

Local artist Kevin Rogers pinpointed 1 pm as the time of day the 'S' was carved. I take his word for it. Kevin is the creator of the John McDouall Stuart Heritage Trail meandering around the edge of Daly Waters, and I'm following the paths which are dotted with billboards covered in maps, diagrams, journal excerpts and Kevin's personal opinions, such as the following:

> **What is amazing is that Stuart and his party would scout in all directions for up to forty kilometres looking for water. He would never move on unless he found a supply for his men and the horses. It is hard to imagine anyone doing something like this in the 21st Century, but these men were made of the right stuff and could endure such insurmountable hardships with the barest of necessities.**
>
> **... Stuart spent 22 days in the region sending scouts**

out in a northerly direction in search of suitable water supplies. This shows the patience and respect he had for this country, unlike the Burke and Wills expedition which just kept on blazing a trail northward without first establishing water supplies.

Keith writes about sinkholes so deep a complete horse and carriage could disappear. This must be the deep holes McDouall Stuart wrote about when he feared he and his men were in danger of getting their necks broken. He wrote in his journal:

This would be a fearful country for anyone to be lost in, as there is nothing to guide them, and one cannot see more than 300 yards around, the gum trees are so thick, and the small belts of lancewood make it very deceptive.

Should anyone be so unfortunate as to be lost, it would be quite impossible to find them again—it would be imprudent to search for them, for by so doing they would run the risk of being lost also.

McDouall Stuart was stuck here for three weeks. 'I feel this heavy work much more than I did the journey of last year; so much of it is beginning to tell upon me,' he wrote. 'I feel my capability of endurance beginning to give way.'

'*Thon water bags.*' I was wondering when he was going to mention them. '*There wis me placing such dependence on them fer carryin us through. Where do I start? Fillin up took too long and the first time we tried them, they leaked out. Half empty after only twenty miles.*

'I ended up usin them tae gie the horses a wee drink at night and in the mornin, but that wis the limit.'

I'm enjoying the air-con in the car as we drive off and I offer to share my water bottle with the face-misting nozzle.

THIRTY-FIVE

NEVER NEVER

McDouall Stuart needn't have worried about the useless water bags. Too much water was the downside when he and his men headed east into the Big River Country. This is the Top End with its lush greenery, palm trees and tropical flowering plants, as different as it gets from the desert landscapes and climate they'd become used to in the Centre.

He was initially thrilled with the Roper River's catchment of creeks and channels, branches and tributaries, spreading out over 80,000 square kilometres. 'This is certainly the finest country I have seen in Australia', he noted in his journal, echoing Gregory's views a few years earlier that this region was perfect for pastoral settlement.

'You finally joined the dots,' I comment. McDouall Stuart is a strange one. He'd crossed over where his fellow explorer passed by, something he was aiming for, and hardly gave it a mention in his journal. He'd done the same a couple of years before, when he crossed the South Australian border, becoming the first white man to do so. It's like once he achieved something, it didn't matter anymore. But it mattered to the South Australian government. Less than a year after this final expedition, the Northern Territory was annexed to SA, and the map of Australia's state borders changed once more.

'And how are you goin' wi' yer ain journal, hen? Keepin a guid record are ye?'

Oh dear, he's right. I'm so tired at the end of each day, driving long distances, setting up and packing up camp each day. Thankfully, I've got hundreds of photos to remind me later where I've been and what I've seen.

Paradise is what I used to call the Top End during the mid-year dry season when I worked up here at the turn of the twenty-first century. You could plan a barbecue without fear of rain. At the moment, it's the hint of dampness I smell, a signal of the build-up before the wet season sets in.

While the Dry is paradise, and lightning storms in the Wet are thrilling, the build-up is an in-betweeny soggy state. It's a lethargy-inducing mixture of heat and humidity lasting weeks on end, when the rain threatens to pelt down but never does, leaving you perpetually wanting. The air is so muggy you feel you're snorting warm tea. A work colleague and I had an idea for T-shirts with the slogan 'Blame it on the Build-up', something to point to in explanation for those moments when you go troppo or merely have a hissy-fit. Pity we were so lethargic we never had the oomph to put that money-making scheme into action.

It's now October, the end of the tourist season, and I can take my pick of where to set up in the Elsey National Park campground where a lone campervan offers the only potential company. I saunter over to the river to check out the sign with thumbnail drawings of people canoeing, picnicking, fishing and swimming. Lead me to the water. Then I read the sign next to it. 'Crocodiles and dangerous currents may occur at this time and can cause injury or death.' Not even a toe dip is allowed. I settle for a walk to look from the banks at those cooling cascades and tumbling waterfalls.

Skinny tree trunks no thicker than a branch have twists and turns like they've had a lifetime following the sun, while others stretch straight and tall, perfect for those telegraph poles that McDouall Stuart was instructed to search for. I capture sparkling light patterns through trees on the river banks and snap close-up shots of branches mirrored below the surface of the royal blue waters, turning back only when my camera battery runs flat.

The trees that shone on the way in are now in the shade and the palms spreading their fronds over the surface of the water are creating menacing shadows. The still waters now hint of solemn secrets and I look for the tell-tale sign of a slithery slide from the sandy riverbank into the water. I'm glad it's not nightfall when the ominous sign of crocodiles would be two red eyes on the surface of the water.

'Not settling... going for a dip,' the young mum from the campervan mumbles when we meet at the shower block, jostling her crying baby on her hip. I break the news and frustration drips off her as she walks off. Her baby's cry soars to new heights. I can't hear what she's saying to hubby back at their campervan as he turns his back, fussing over something inconsequential. But I can hear the tone as they play out a version of Unhappy Families. Mum: Tired. Exhausted. Dad: Bewildered. Walking away. 'Blame it on the Build-up.'

It was June, the dry season, when McDouall Stuart passed this way. Perfect timing—if you ignore the fact that he was, by now, supposed to be back home in Adelaide. He's still 500 kilometres or so away from the tip of the Top End.

He did have his problems. That wonderful Roper River was proving hard to cross and one horse after another was getting into deep water, the men struggling to free them. One horse just couldn't get out, the gum tree branches having fallen both above and below him. He was completely fixed. 'We endeavoured to get him out but it got so dark, we could not see him,' McDouall Stuart wrote in his journal. 'The rope we were pulling him out by broke, he got his head underwater, and was drowned in a moment.'

His entry two days later: 'We are all enjoying a delightful change of fresh meat from dry. It is a great treat and the horse eats remarkably well'. Even though he had a soft spot for those horses, he's a thrifty man who doesn't let an opportunity pass him by. Auld also felt this incident worth recording, pointing out that they had three meals of freshly killed meat, complete with native cucumbers as a relish.

Tonight, I'm dining on carrots and apples, what people feed to horses I believe. Once again, I can't be bothered lighting a fire, or even setting up the burner.

It was clear that McDouall Stuart was wary of the locals, who were making it clear that this was their country. 'I wish it would rain and cause the grass to become green, so as to stop them burning,' he complained. The region was 'thickly inhabited' he wrote and they were 'not to be trusted'. The locals felt the same way about him, running along the opposite river bank, lighting fires as they went, 'setting up a fearful yelling and squalling and running off as fast as they could' whenever they came in close contact. On one occasion, McDouall Stuart struck a match to light his pipe and the natives gave a yell and cleared off.

Sometimes the groups did approach each other. Out of curiosity. One old man was fair taken with a fish hook stuck in Kekwick's hat, and was thrilled to receive it as a gift. On another occasion, it was Kekwick who was curious to see what fish one fellow, armed with spears, had in his very full bags.

The flies during the day were a nuisance (oh how I could sympathise with that one) and the mosquitoes at night were terrible. Stephen King recorded later: 'Our hands, wrists, necks and feet were all swollen and blistered from their bites. I do believe they taught us to swear very fluently at times.' No doubt they did so under their breath, considering McDouall Stuart's views on that matter.

'Pity you yersel' huvnae considered ma views on that matter, hen.' I say nothing. The two of us get on famously if we ignore each other at moments such as these.

Elsey National Park borders on one side with Elsey Station, one of the first pastoral properties in the Northern Territory, and immortalised in the book *We of the Never Never*. On the other side of the park is the tourist spot of Mataranka, capitalising big time on the book and a movie of the same name; thundering horses' hooves and all.

The book, written by Jeannie Gunn after living at the station for one year, and published more than a hundred years ago, was taught in schools throughout Australia, on and off, and has never been out of print. Her romanticised story helped to set off what has become an everlasting love affair with the Outback: the first of many flawed portrayals of a Territory history that, in reality, never existed. For one, the pioneering outback life was far from idyllic. Two, the pastoral industry was never prosperous. And three, the promotion of the book, labelled a classic, and a 'fresh, affectionate and minutely observed account of tropical outback life', sidesteps the truth about frontier violence.

In an ABC radio Background Briefing report in 1999, journalist Lorena Allam points out that, in the book's chapter called 'A Nigger Hunt', later renamed 'A Surprise Party', Jeannie, her husband, and a group of stockmen, ride out with rifles in search of 'blacks' interfering with the cattle.

'Surprise party' is one of those euphemisms often used at the time in newspapers, reports and conversation along with 'teach a lesson', 'a picnic with the natives', 'punitive expedition' and 'dispersing'.

In a reflective paragraph, removed in some editions, Jeannie suggests the killings had stopped by the time she arrived:

> A black fellow kills cattle because he is hungry and must be fed with food... And until the long arm of the law interfered, white men killed the black fellow because they were hungry with a hunger that must be fed with gold...And yet men speak of the superiority of the white race...

But Lorena Allam points out: 'There is evidence that shootings were going on at Elsey Station while Jeannie Gunn was there.' This is revealed by Jack McLeod, immortalised as the Quiet Stockman in the book, during a 1958 radio interview where he recalls taking part in one himself.

Executive Producer of the 1980s film *We of the Never Never,* generally panned for being sanitised, sentimental and stereotypical, is radio broadcaster Phillip Adams, who describes the movie as a Disneyfied distortion of a distortion.

'I dimly thought it might be useful to make a film (from the book) which gave some recognition to Aborigines in Australia,' he said. 'I remember the book as being progressive in attitudes for its time. When I belatedly read the damn thing, my blood froze because I realised that it was infinitely patronising.

'I regret that it (the movie) was ever made and of course, wouldn't touch it with a barge pole were the proposal to be brought to me today.'

The movie remains on a constant loop in a replica of the original homestead at Mataranka, giving tourists a full dollop of fake history. Near the replica homestead is an artificial spring kept at a constant 34 degrees Celsius, to make the tourists more comfortable, and a talking termite mound is within walking distance. How much falseness can one tourist spot conjure up?

Linguist and anthropologist Francesca Merlan describes the period after McDouall Stuart came through, beginning with the construction of the Overland Telegraph Line, as one of guerrilla warfare with battles for control of land and water along the Roper River intensifying as more and more country was taken up by pastoralists.

Irishman Abraham Wallace was the first white settler to take up the lease for Elsey Station in 1882, part of the pastoral boom at the Top End, administered by the colonial government in Adelaide.

The early settlers saw violence towards Aboriginal people as necessary. 'Making people quiet' is the phrase the Mangarayi use to describe the treatment of their ancestors, considered a 'problem' and a menace to stock.

The Mangarayi remained loyal to their country as it changed hands several times and they watched the land deteriorate as cattle upset the ecological balance—devouring the grasses, hardening the ground with their hooves and taking over the waterholes.

The power structure of stations such as this one, run like mini fiefdoms, devastated the social system that had existed for countless generations among the First Nations peoples. From the turn of the twentieth century, Mangarayi worked on Elsey Station for rations and shelter, and were treated as though they were the possession of the station owners. As the north became more settled local Aboriginal groups shifted from violent resistance to begrudging acceptance of the situation, for they had at least some access to their country.

It wasn't until the last quarter of the twentieth century, not that long ago, that Aboriginal land rights were recognised in the Northern Territory under the *Land Rights Act*. In 1991, the pastoral industry was in decline and the lease for Elsey Station was purchased on behalf of the local Mangarayi people, making it claimable under the legislation. And so, the long, laborious process began.

In 2000, the title deeds of the property were handed over to the traditional owners of the area, a section of the land they'd looked after for thousands of years. It took nearly nine years of legal to-ing and fro-ing for the Mangarayi to prove their continual connection with their ancestral homelands, achieved despite the harshest aspects of colonisation threatening those bonds.

The wheel has come full circle, said Henric Nicholas, a direct descendant of Abraham Wallace, the first leaseholder of the land, speaking at the formal hand-over.

'The traditional owners, once again, have total rights to move around as they wish, to fish and hunt, and live wherever they want,' he said. 'These rights were laid down in that very first contract and every one since—but ignored.'

THIRTY-SIX

MY TURN TO WHOOP

I'm sitting in the Bark Hut Inn on the Arnhem Highway, contemplating a piece of buffalo pie. The Bark Hut is my last chance for fuel, food and water before I head for my final destination, Point Stuart, with Chambers Bay on one side, Finke Bay on the other.

McDouall Stuart made his one major map-reading blunder up here in the Top End, not having a chronometer to accurately calculate longitude, and getting his rivers mixed up. What a difference half a degree on a map can make.

It's the end of the tourist season and the time of year when things can get boggy. An error in longitude would have been the least of my worries in this terrain, which is why I had headed up the Stuart Highway instead of following his tracks. I planned to go straight to Darwin. Then I spotted Arnhem Highway, off to the right, and after a quick shuffle through my papers I realised that I could reach the very spot where McDouall Stuart reached the coast. I've just crossed the Adelaide River near Humpty Doo (my all-favourite name for an Australian town) and the Mary River, the two stretches of water that had him stumped.

As I order that piece of pie and pull out my map, I'm informed that the young man in the kitchen is the one to help me with directions to Point Stuart. His credentials appear to rest on a love of fishing in the rivers and waterholes in the general area I'm pointing to.

I've been on the road for one month, zig-zagged 6,000 kilometres from the Mediterranean climate of Adelaide, through the arid zone of Central Australia and now getting a little steamed up in the tropical Top End. I'm not ready for the journey to end. And it's not because I've already drunk the piccolo of sparkling wine meant for the occasion.

McDouall Stuart and his men, however, were more than ready for the journey to end. For days now, they'd pushed forward, each detour, each obstacle frustrating their progress after leaving the Roper River. The horses were struggling, weakened by worms and hampered by worn-out horseshoes as they sank into the boggy marshes, buckling under their heavy loads. Did I mention horseshoes? The men were exhausted; tormented by flies during the day and pestered by mozzies during sleepless nights.

'I've never ventured up to Point Stuart myself,' says the young lad from the kitchen, with a look that suggests why would anyone want to.

He's more interested in warning me about crocodiles, relishing recounting a story he finds hilarious, about three men in a tinny at night, surrounded by a dozen pairs of red eyes, circling closer and closer. The Mary River, which I'll be following, has fifteen crocodiles per kilometre, the highest density of any river in Australia.

'Best of luck... don't get lost.' If I was expecting a pipe-band fanfare to see me off, it doesn't happen. I leave with a full tank of fuel, replenished water supply, a bottle of red to replace that piccolo I drank back near Chambers Pillar and enough food to last a couple of days. Fingers crossed.

As instructed, I take the first sealed road on the left. It doesn't stay sealed for long, becoming a rutted and winding track, with fallen tree branches and creeping undergrowth threatening to take over. A faded arrow, stuck on a tree trunk, points left, but the track looks like it may peter out. So, fuck it, I go right, ignoring the 'Private Property' sign.

Bouncing along the track, my vision is of a line of stunted stringy barks. I'll roll down the window, smell the sea air, and hear the whisper of the wash lapping the shoreline. Just as McDouall Stuart had described. I'll break through the dense scrub lining the beach and whoop 'the sea, the sea'. Just like the young Francis Thring who got the honour of being the first on the beach.

But it's not the sea I see as I round the corner. Before me is a clearing, ringed by a scatter of demountable buildings. I've arrived at the private Stuart's Tree Fishing Camp. Like the trespasser I am, I hug the trees around the outside edge, heading for my getaway—a narrow track I've spotted at the far side. But it leads nowhere and I'm soon back to the clearing. I spot a figure on the front veranda of the largest building, with a walking stick, or could it be a shotgun, in hand. Still as a statue, but rotating, like a board game I, for some reason, recall from my childhood, and it's following my circular path. It's like she knew I'd eventually come to heel. It's a shotgun she's holding.

Composing my excuses, I touch Bad-Ass Betsie's chassis for moral support, leaving the door open to show I'm not intending to stay. I've never been so pleased that Betsie isn't a campervan or mobile home, or anything suggesting I was deliberately planning to trespass and linger.

'Visitors usually arrive by sea.' She nods over my shoulder to Betsie who's doing a bit of huffing and puffing. I did come to a rather sudden halt. 'And never this late in the year.' I count nine dogs wrapping themselves around her legs, a very tame wallaby is watching from the sidelines, a gaggle of geese is waddling towards us and a very annoying baby corella in a cage is screeching for attention.

'Fancy a cuppa?' Heather Rose, caretaker of the Fishing Camp, is looking for human company.

As Heather Rose boils the kettle, I look around. An open folder is on the dining table. I recognise that face. A map is on the fridge and a poster is on the back of the bathroom door. I'm in luck. Another McDouall Stuart fan.

Heather Rose is my personal guide, pointing out the tracks of buffalo, wallaby, goanna and snake and making sure I avoid crashing through huge, glistening spider webs wrapped around and between the massive paperbark and banyan trees. She points out the hollows where the buffalo wallow in the mud. We squish through the mangroves, skirt a mighty Aboriginal midden where it looks like thousands of years' worth of pipi and longbum shells are piled up high, and walk along the beach of gritty sand edging the scummy water.

I imagine McDouall Stuart standing at the same water's edge in 1862, fully clothed except for his bare feet, his ankles like two bleached bones sticking through the soft, blue mud. He stooped down and, favouring his good hand, splashed his face as he'd promised South Australia's Governor MacDonnell he'd do. So do it he did. I've no intention of dipping my toes in the water: not with that crocodile cage back in the mangroves.

McDouall Stuart had his initials carved into a tree not far from the beach and the following morning, after packing for the journey home, he had the Union Jack, sewn especially for this occasion by Chambers' daughter Elizabeth, fixed to the highest branch of the highest tree where they'd camped overnight. Each expeditioner signed a commemorative message which was sealed in an air-tight canister and buried close by. Three cheers for the Queen, then three cheers for the Prince of Wales. It was the 25th of July in the year 1862 and, as John Bentham Neales prophesied, it was indeed nine months to the day since the farewell luncheon in Adelaide.

Auld was volunteered by the others to suggest they might indulge in a small celebration. 'A cup of tea, some broth, maybe?' Auld proposed. There would be none of that, said McDouall Stuart, and it would be a few days before Auld got back into his good books.

Heather Rose and I crash through the undergrowth and there it is. We're at the McDouall Stuart Memorial, the cairn that's replaced the tree with the carved initials.

My turn to whoop.

EPILOGUE

'*Sit doon. Let me tell ye a story.*'

'Fuck. I thought you'd gone.'

'*Language young lady.*'

'What d'you expect? I thought we were finished. Story over.'

'*C'mon, you can't end it there. One of ma achievements wis gettin'
up to the top, the ither ane was gettin back doon withoot losing a man.
You've got to finish it off.*'

'Oh. Okay. Will I write about you having to be carried on a
stretcher between two horses for the last bit?'

'*Before that. I huv a bone tae pick wi' ye. You made me sound like a
tyrant, not lettin' them celebrate.*'

A tyrant? No. Not a tyrant. A martinet, maybe.'

'*I didnae think we'd make it back.*'

'You nearly didn't.' I'd seen the sketch that King drew of the
stretcher. Some called it an ambulance.

'*That's why I wouldnae let them waste any burgoo on a toast. We
needed every morsel we had to get us all hame.*

'*Remember you wrote aboot all those incidents at Mount Hay? Well,
you should have seen me there on the way doon. If the natives had wanted,
they could have picked me off nae bother.*'

'Anyway, you got back down, so did all the men and a good percentage of the horses.'

'*Aye, we did that.*'

On arrival at Moolooloo, McDouall Stuart was told that Chambers
had died. Governor MacDonnell, one of his biggest supporters, had

left the country. Finke, his other benefactor, died not long after.

'And there's the spectre of Burke and Wills again, with their bodies carted through Adelaide only days before you got back into town. You couldn't exactly holler with glee at being successful. A bit too soon for that.'

'*Aye. But what a demonstration they put on a few weeks later.*' I'd read about that.

A public holiday was proclaimed, tartan bunting fluttered, copious speeches were made, cheers were hollered loud and often, and toasts were raised.

But once again Burke and Wills stole the limelight.

'You were in their shadows right to the end, weren't you? Fancy their state funeral, Australia's first, being held in Melbourne on the very same day as your Great Demonstration.

'*Och. Who cares. It wisnae ma funeral.*'

McDouall Stuart was granted his £2000. Later, it was decreed that this money should go into a trust, and he only ever got £200 of it. He also gave up the lease on that piece of land at Stuart Creek, never to get time to enjoy it, a reward hard won and then forfeited.

He sailed back to Britain a couple of years later to oversee the publication of his journals and died two years after that.

That tree with his inscription wasn't discovered until twenty years later, when the doubters finally believed McDouall Stuart had reached the coast. Just as well it was spotted when it was, as that tree was soon gone altogether. Depending on which report you read, it either died, was vandalised or got burnt in a bushfire.

So this is it, is it? End of the trip, end of our cosy wee chats.'

'S'pose so.' I'd no intention of driving back down. 'It's been nice knowing you Mr Stuart.'

'And you tae Ms Cadden.'

It's off into Darwin for me, Gulumoerrgin country to the Larrakia people, where I'll be putting Bad-Ass Betsie on the Ghan back to Adelaide. I'm off to Scotland, where I plan to stay a couple of nights in the house in Dysart where McDouall Stuart was born.

'Let's go Betsie.' I jump behind the steering wheel and we're off. 'Our last jaunt together.'

'I might just join you in Dysart. What have you got to say aboot that?'

'FU-U-U-U-CK'

EXPLORERS' ROUTES

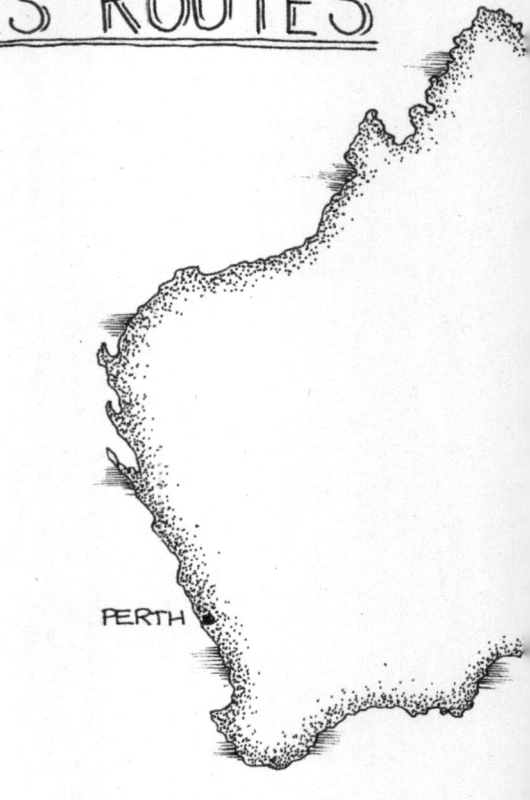

PERTH

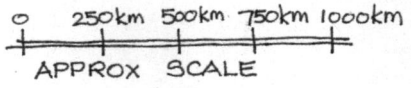

0   250km 500km 750km 1000km

APPROX SCALE

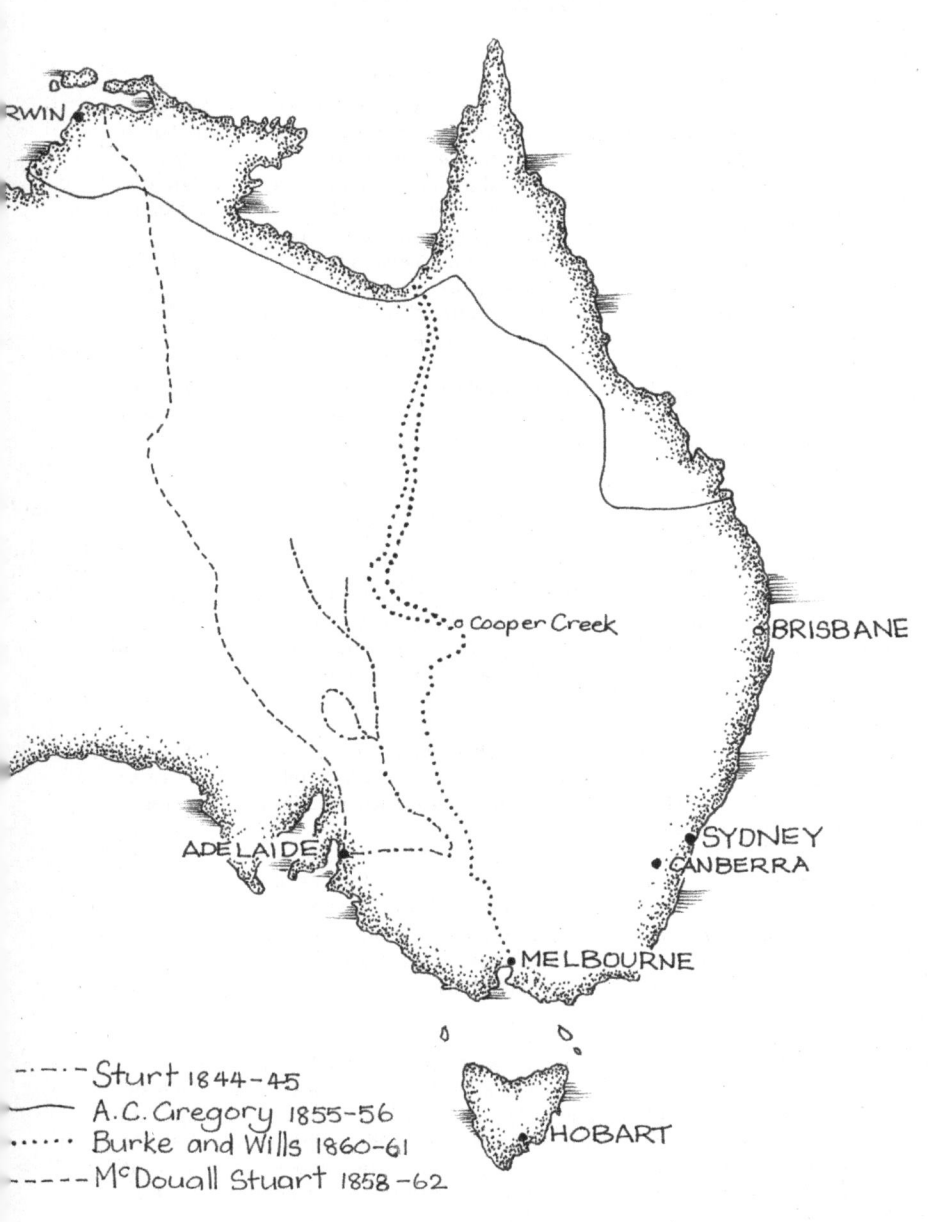

RWIN

o Cooper Creek

BRISBANE

SYDNEY
CANBERRA

ADELAIDE

MELBOURNE

HOBART

-----Sturt 1844-45
——— A.C. Gregory 1855-56
••••• Burke and Wills 1860-61
-----McDouall Stuart 1858-62

AUTHOR'S NOTES

Stuart's Pea (page 14)
The Sturt Desert Pea *(Swainsona formosa)*, named after explorer Charles Sturt, is a striking creeping vine with vibrant red flowers. It is the floral emblem for South Australia, found on tea towels and mugs, badges and bookmarks, at one time on a postage stamp. Not so widely known and with understated elegance is Stuart's Pea, with its small yellow flowers. Both are found at the Adelaide Botanic Gardens. I first learned about Stuart's Pea when reading an article by Dennis Hall 'Introducing a Desert' in the book *First Across the Simpson Desert* by E.A. (Ted) Colson. My interest was piqued. Because I didn't know the botanical name of the plant, I couldn't track it down, until I met Diarshul Sandhu, a volunteer guide in the Gardens, who knew about Stuart's Pea, took me to see it, and told me its botanical name *(Crotalaria cunninghamii)*.
The plant, I have to say, did not look particularly impressive. Until I looked closely. Stuart's Pea has become my logo to remind me that first impressions are not to be trusted and that what you see and learn depends on where you are and how you look at it.
See my blog for the full story:
https://alongthoselines15.wordpress.com/2023/08/08/back-again

Proclamation Day in SA (page 24)
That scene under the gum tree in Glenelg, depicted in the Art Gallery painting, is re-enacted every year, and the citizens of Adelaide get a day off work to commemorate the occasion. Most don't know why or what the Proclamation is all about but it's a good excuse for a party, just like the original knees-up.

Letters Patent (page 25)
The exhibition 'Tiati Wangkanthi Kumangka' (Truth Telling Together) at the Bay Discovery Centre in Glenelg, a short tram ride from Adelaide, tells the true history of South Australia and features the Letters Patent which includes the following paragraph, largely ignored:

> 'Provided always, that nothing in these our letters patent contained shall affect or be construed to affect the rights of any Aboriginal Natives of the said Province to the actual occupation or enjoyment in their own Persons or in the Persons of their Descendants of any Lands therein now actually occupied or enjoyed by such Natives.'

Kirkcaldy (page 27)
The South Australian Company, aiming to attract immigrants to this new 'province', must have had a very active marketing push in the Kingdom of Fife in Scotland. Apart from explorer John McDouall Stuart, the list of folk from Kirkcaldy and surrounding towns who sailed over includes the Elder siblings (Elders Real Estate and Elder Conservatorium of Music), Peter Waite (Waite Research Institute and Urrbrae Agricultural High School) and Sir Walter Hughes, knighted in 1880 (a founder of the University of Adelaide).

Through Walter Hughes, I also like to add AFL footballer and Australian of the Year, Adam Goodes, to the list. Hughes, from Pittenweem along the coast from Kirkcaldy, is his ancestor, a fact Goodes discovered during the filming of an episode of the ABC TV program 'Who Do You Think You Are'. Hughes and an Aboriginal woman in Moonta had a relationship. Their son, John Sansbury, had a son, Edward, who married Jessie Johnson, an Adnyamathanha woman and Adam Goodes' great-great-grandmother.
(Series 6, Episode 6 Aired 12 August 2014)

The Southern Cross (page 47)
This is now on the Australian flag. The design has been linked to Christianity and colonialism; to working-class ideals and neo-Nazi leanings. For Aboriginal people, one interpretation focuses not so much on the stars but the spaces in between, where the Spirit Emu lives.

National anthem (page 48)
One word was changed in the Australian national anthem in 2021. The symbolic tweak is from 'young and free' to 'one and free', belatedly acknowledging that, far from being young, the First Nations peoples of Australia arrived here 60,000 years ago and counting, making it the oldest surviving culture in the world.
Missed a chance not to do something at the same time about that word 'girt', methinks.

Census (page 49)
Far from dying out, which was the belief in Australia for many a decade, nearly one million people in the 2021 Census identified as being of Aboriginal or Torres Strait Islander descent.

White Australia Policy (page 49)
The origins of the *Immigration Restriction Act 1901*, commonly called the 'White Australia Policy', can be traced to the 1850s and white miners' concerns about Chinese diggers. The policy describes Australia's approach to immigration from Federation until the latter part of the 20th century, which favoured applicants from certain countries and upheld the view that Australia was 'an outpost of the British race'.

Oodnadatta water (page 124)
Good clean water has been provided in Oodnadatta. Finally.
When I was there in 2015, I was advised not to drink it without boiling and I'm glad I took that advice. An Indaily report in August 2023 confirmed that the town's residents had been charged for undrinkable water for decades. Not only undrinkable but also a potential source of the parasite, *Naegleria fowleri*. Now they can safely drink from their taps after a desalination plant began operating in the remote South Australian town.

Voice to Parliament (page 179)
'First Nations people find they are not spoken to, rather they are spoken about,' was a pertinent point in a presentation at the CRANAplus conference in Alice Springs in 2015. The presentation was written by Pat Anderson AO, an Alyawarre woman, human rights advocate and known nationally and internationally as a powerful advocate for the health of Australia's First Peoples.
More recently, Pat Anderson was co-chair of the Referendum Council and one of

the architects of the Uluru Statement from the Heart. She campaigned for every Australian to understand that the heart of the proposal for an Indigenous Voice to Parliament would empower Indigenous people to have a say on matters that affect them.

Australians overwhelmingly voted 'No' for an Indigenous Voice to Parliament on October 14, 2023. The referendum failed. But data shows the majority of Indigenous communities voted 'yes'.

Great-great-grandson of Waterhouse (page 240)
Dr Andy Thomas, born in Adelaide, became the first Australian-born space astronaut when he flew his first flight in space on *Endeavour* in May 1996. He completed four flights, logging more than 177 days in space, before retiring in 2014. He is the great-great-grandson of Frederick George Waterhouse, who was on McDouall Stuart's final expedition in 1861-1862, and he took expedition artefacts with him on his first space flight. Dr Thomas is the patron of the John McDouall Stuart Society.

SCOTTISH GLOSSARY

Scots is one of the three indigenous languages spoken in Scotland alongside English and Scottish Gaelic. Scots received official recognition as a language in its own right by the Scottish Government in 2015, followed by recognition by the UK government and the European Union. Scots remains an essential part of Scotland, its history, culture and identity.

blether
To have a blether is to chat. To be a blether means you talk too much

canny
A Scottish word, used in many English-speaking countries, meaning to be shrewd or careful

corbie
Either a raven or a crow

doddle
It's a doddle, means it's very easy

dug
Some say a dug is your best friend – a dog in other words.

fankle
To get yourself in a fankle: to get in a muddle, physically or mentally.

feart
To be feart is to be afraid. Where I come from 'a fearty gowk' is used by children to mock their pal for being scared of something inoffensive.

footer
As a verb, it means to fidget or fumble. As a noun; 'He's an auld footer' is someone who is exasperating. Not to be confused with First Footing. (See Hogmanay)

glaikit
Some folk ask if glaikit is a real word – but you'd be a bit glaikit to think that. Meaning: foolish.

haver
To haver is to talk nonsense, as heard in the Proclaimers song 'I'm gonna be (500 miles).'

hen
A term of endearment for a woman, usually young

Hogmanay
New Year's Eve in Scotland. The first person to enter a house after midnight (the first footer) should be a tall, dark-haired man bringing gifts: such as coal, whisky or black bun.

jalouse
To surmise

kent
Past tense of 'ken', meaning to know

scunner
It's a scunner. Something or someone irritating, annoying. 'It's a right scunner.'

shoogly
Shaky, unsteady. If someone's job is insecure, you could say "It's on a shoogly peg.'

spyug
A common sparrow. If you think a restaurant meal is very small, you might say: 'It widnae feed a spyug.' There are different spellings.

Yer aff yer heid
This is a more modern Scottish saying. You're off your head – meaning you are a bit daft.
In Australia, it's the same as: 'You're dreaming', a phrase from the movie *The Castle*.
Keep the heid! is another popular phrase. It means: 'Stay calm, don't get upset.'

A few pointers about the Scots language:

Negative forms of some verbs are created by adding 'nae' (or 'na') at end – 'cannae', 'widnae', 'dinnae' etc

Present participles end in 'in' – almost never 'ing', so there is no need for an apostrophe – greetin (crying), slaverin (talking nonsense).

In Scots, you can combine modal verbs so that they are adjacent in a sentence, such as 'He micht kin dae it later' Literally – He might can do it later (No promises!). It's a bit like the double negative, used to emphasise a point. ' Gonna no dae that! Meaning: 'Definitely don't do it.'

Scots speakers sometimes use feminine pronouns for some objects to do with nature, such as the weather, and bodies of water. 'She's wairm oot the day' – referring to the warm temperature outside.

Scots uses older, short vowel sounds in words like 'hoose', 'mooth' and 'coo' (like Norwegian) instead of 'house', 'mouth' and 'cow' (like English).

Scots language and culture courses are jointly produced by The Open University in Scotland, the Open University's School of Languages and Applied Linguistics and Education Scotland.

FURTHER READING

Here's a list of the primary and secondary sources I used in my research; details of references within the book; plus books, reports, newspaper articles, websites and YouTube videos that I think you may find interesting.

PART ONE

Magazine
Australian Geographic magazine, No 11 July-September 1988 'Exploring the Stuart Highway and the Oodnadatta Track'. *I still have that magazine, which started this project, complete with dog-eared corners and scribbles.*

Article
Anderson, Ken 'John McDouall Stuart and the Great North Road: http://www.aicomos.com/wp-content/uploads/John-McDouall-Stuart-and-the-great-North-Road.pdf

Books: Early readings; explorers, first settlers
Bailey, John *Mr Stuart's Track: The Forgotten Life of Australia's Greatest Explorer*, Pan Macmillan Australia, 2006.

Brock, Peggy and Gara, Tom. (editors) *Colonialism and its Aftermath*, Wakefield Press, 2017.

Collins, Carolyn and Sensziuk, Paul (ed) *Foundational Fictions in South Australian History*, Wakefield Press, 2018.

Kerr, Margaret G *Colonial Dynasty: The Chambers Family of South Australia*, Adelaide: Rigby,1980.

Mattingley, Christobel (ed), Hampton, Ken (co-editor) *Survival in our Own Land*, Melbourne, Australian Scholarly Publishing,1998.

Payton, Philip (ed) *Emigrants & Historians*, Wakefield Press, 2016.

Sendziuk, Paul and Foster, Robert *A History of South Australia*, Cambridge University Press, 2018.

Webster, Mona, S. *John McDouall Stuart*, Melbourne University Press, 1958.

Journals
McDouall Stuart's journals from his six expeditions are available on the John McDouall Stuart Society website
http://johnmcdouallstuart.org.au

YouTube
Kaurna – plants, animals, seasons, culture and history https://www.youtube.com/watch?v=REhdXAJGZVI

Websites
'Exploring the Stuart Highway and the Oodnadatta Track' *is available online at* www.exploringaustralia.com.au
  Bay Discovery Centre
  SA History Hub

Museum of Australian Democracy
Professional Historians Association of SA
South Australian Native Title Services

PART TWO

Books: My early years' readings in Australia

Blainey, Geoffrey *The Tyranny of Distance*, Pan Macmillan, 1967. *His view is that distance and isolation have shaped Australia's identity.*

Horne, Donald *The Lucky Country*, Penguin, 1964. *The lucky country is a phrase often used to give Australia a pat on the back but Horne intended the title of his book to be a criticism.*

Morgan, Sally *My Place*, Fremantle Press, 1987. *It reveals how Aboriginal people hid their heritage to avoid discrimination.*

Pike, Douglas, *Paradise of Dissent 1829-1957*, Melbourne University Press, 1967. *The First Nations peoples of Australia hardly rate a mention in this book, published when assimilation was at its zenith.*

Books: Belonging

Krichauff, Skye *Memory, Places and Aboriginal-Settler History*, Anthem Press, 2017.

Read, Peter, *Belonging – Australians, Place and Aboriginal Ownership*, Cambridge University Press, 2000.

Smith, Anne-Marie (ed) *Culture is... Australian Stories Across Cultures*, for the Multicultural Writers Association of Australia, Wakefield Press, 2008.

Books: Flinders Ranges

Barker, Sue et al (eds) *Explore the Flinders Ranges*, Royal Geographical Society of SA Inc. Adelaide, 2014.

Bruce, Robert *Reminiscences of an old squatter*, Adelaide, W. K. Thomas, 1902 *Available through the State Library of South Australia.*

Mincham, Hans *The Story of the Flinders Ranges*, Rigby, Adelaide, 1964.

Neal, Louise, *An earl's son – The letters of Hugh Proby*, Self-published, 1987.

Sheldrick, Janis *Nature's Line: George Goyder*, Wakefield Press, 2013.

Books: Cultural Heritage

Royal Geographical Society of Australasia *Reminiscences of Johnson Frederick Hayward*, Proceedings of the RGSA, South Australian Branch vol. 29 1927-28, pp. 79-170, 2019. *It has some interesting insights into his experiences.*
www.samemory.sa.gov.au/site/page.cfm?c=5287
Many people living in the Flinders Ranges have researched books about their families, the stations they live on, the towns they are close to and the region as a whole. If you travel through the area, you are likely to come across them where you buy a coffee or book a campsite. They include:

Carrieton Centenary Book Committee *Carrieton in the Gum Creek Country*, South Australia, 1978.

Ragless, Margaret *Dust Storms in China Teacups*, Ragless Reunion Committee, Investigator Press, 1988.

Books: Aboriginal agricultural practices

Gammage, Bill *The Biggest Estate on Earth: How Aborigines Made Australia* Allen & Unwin 2012. *Historian Bill Gammage exploded the myth that pre-settlement*

Australia was an untamed wilderness.
Pascoe, Bruce *Dark Emu: Black Seeds: Agriculture or Accident*, Magabala Books, 2014. *In this, he argues for a reconsideration of the 'hunter-gatherer' tag for pre-colonial Aboriginal Australians and attempts to rebut the colonial myths that have worked to justify dispossession. Check YouTube for documentaries on The Dark Emu Story.*

Film: First encounters and frontier violence
McKinnon, Malcolm & Thomas, Jared (Dirs) 'Close to the Bone', IMDb, 2022. *A documentary and collaborative project for ABCTV and the South Australian Museum*

Project: Digital mapping
Krichauff, Skye 'Reconciling with the Frontier' *A University of Adelaide project which includes a digital map to share sites in South Australia where colonial violence took place against Aboriginal people in the 1800s. Here's a podcast: https://www.nativetitlesa.org/ mapping-colonial-violence-against-aboriginal-people-in-sa/*

Book: Native title
Clarke, John *Still the Two*, Text Publishing, 1997. *This includes the transcript of a brilliant skit about native title, using a squabble over a pen to illustrate the stance of John Howard, then prime minister of Australia. Do try to find it. You won't be disappointed. The Central Land Council website www.clc.org.au has a plain language explanation of Native Title.*

Article: Trading routes and red ochre
Mulvaney, John '...these Aboriginal lines of travel', *Historic Environment, 16.2 pp 4-7, 2002. John Mulvaney, the first qualified archaeologist to focus his work on Australia, writes here about the social and economic role of Dreaming tracks and their environmental relevance. https://www.aicomos.com/wp-content/uploads/these-Aboriginal-lines-of-travel.pdf*

Book: Dreaming stories
Brock, Peggy *Yura and Udnyu*, Wakefield Press, 2019.
Tunbridge, Dorothy, in association with the Nepabunna Aboriginal School and the Adnyamathanha people of the Flinders Ranges, South Australia *Flinders Ranges Dreaming*, Aboriginal Studies Press, 1988. *These are considered prime resources.*

Website: Farina
The website for the Farina restoration group is full of information, and you may be tempted to do some volunteering. https://farinarestoration.com

Article: White Australia Policy
Irving, Helen 'One Hundred Years of (Almost) Solitude: the Evolution of Australian Citizenship', *Paper presented at the Senate Occasional Lecture Series at Parliament House*, 22 June 2001. www.aph.gov.au/binaries/senate/pubs/pops/pop37/irving.pdf

PART THREE

Book: Oodnadatta Track
Dodd, Reg and McKinnon, Malcolm *Talking Sideways: stories and conversations from Finniss Springs*, University of Queensland Press, 2019.
I met Arabana elder Reg Dodd OAM in Marree when I returned to the region to do a

volunteer stint at Farina some years after my road trip.

Website: Mound Springs
Check out the website for Friends of Mound Springs
https://www.friendsofmoundsprings.org.au/

Articles
Wabma Kadarbu Conservation Park Management Plan
Google this to get the full report.
This is an article in the Sydney Morning Herald
smh.com.au/environment/sustainability/south-australia-s-disappearing-springs-raise-questions-for-miner-bhp-20201117-p56f6m.html

Article: Anna Creek station
McAuley, Ian 'The Bush – myths and reality', *Dissent*, 29 Autumn/Winter, 2009.
It is available here:
https://www.ianmcauley.com/academic/default.htm

PART FOUR

Booklet: Stuart, Kekwick and Head
Moore, Rick *Stuart, Kekwick & Head – the character of the men*, John McDouall Stuart
Society, 2010. *Giving insight into the three men who undertook the fourth expedition.*

Books: The bush
Cross, Jack *Great Central State: the foundation of the Northern Territory*, Wakefield
Press, 2011.
Watson, Don, *The Bush*, Penguin, 2016.

Article: Witjara-Dalhousie Springs
Colin Harris 'Five Decades of Watching Mound Springs in South Australia'
The Proceedings of the Royal Society of Queensland Journal, 1 January 2020.
Available on the Internet

Article
Australian Financial Review article about BHP
https://www.afr.com/companies/mining/can-a-new-mine-save-bhp-s-loss-making-olympic-dam-20201026-p568sn

Books: Alice Springs
Botanist and author Peter Latz, born at Hermannsburg, a Central Australian mission,
has spent his life raising understanding of arid zone ecosystems. His books include:
Latz, Peter *Bushfires and Bush Tucker – Aboriginal plant use in Central Australia*, IAD
Press, 2018.
Latz, Peter *Blind Moses – Aranda man of high degree*, IAD Press, 2014.
Glenn Morrison, writer, journalist and academic, lives in Alice Springs and has written
extensively about the town and the surrounds. His essay 'No direction home: race and
belonging in Alice Springs' *won the 2013 CDU Essay Award. His books include:*
Morrison, Glenn *Songlines and Fault Lines: epic walks of the Red Centre*, MUP, 2017.
Morrison, Glenn *Writing Home: Walking, Literature and Belonging in Australia's*
Red Centre, MUP, 2017.

Traynor, Stuart *Alice Springs: from singing wire to iconic outback town*, Wakefield Press, 2016.

Television
ABC TV series 8MMM Aboriginal Radio. First aired in 2015, and available on iview, *is a must-watch. The three Ms of the title refers to the three kinds of white people who typically make their way to Aboriginal communities: missionary types, mercenaries in it for the money and misfits.*

The National Film and Sound Archive of Australia, 'A Wire through the Heart', 2007. *This ABC TV documentary is about the construction of the Overland Telegraph Line, which has a segment about John McDouall Stuart.*
https://www.youtube.com/watch?v=TPFSdX9r_Gk

Map: Colonial frontier massacres
Map of colonial frontier massacres in Australia 1788-1930
http://c21ch.newcastle.edu.au/colonialmassacres

Articles: Scottish connection
Cahir, Fred, Inglis, Alison & Beggs-Sunter Anne, (eds) *Scots Under the Southern Cross*, Ballarat Heritage Services, 2014. *It is a collection of essays from speakers at the Scottish Symposium held in Ballarat. Here is a link to an essay by Jason Gibson, Museum Victoria, called 'John McDouall Stuart remembered in Central Australia' which can be downloaded.*
https://www.academia.edu/12602257/
John_McDouall_Stuart_Remembered_in_CentralAustralia

PART FIVE

Books: Burke and Wills
Many books have been written about them including:
Clune, Frank *Dig: The Burke and Wills Saga*, Times House, 1986.
Moorehead, Alan *Cooper's Creek*, Hannah Hamilton, 1965.
Murgatroyd, Sarah *The Dig Tree*, Text Publishing, 2002.

PART SIX

Book: Final expedition
Linn, Rob *Sketching with Stuart.* John McDouall Stuart's 1861-62 expedition seen through the sketches of Stephen King, published by The Friends of the State Library of South Australia Inc. 2017.

Journals
Stuart, J McDouall 'J. McDouall Stuart's Explorations Across the Continent of Australia: 1861–62', Friends of the State Library of South Australia Inc, 2012.
Lawrenson, Elizabeth 'Stephen King Jr (1841-1915)', Pioneers' Association of South Australia (Series); no. 44, 1970.
Auld, William Patrick 'Recollections of McDouall Stuart'. *These were first published individually in the Adelaide Observer in 1891 and 1904 and in the Register, Adelaide in 1910 and 1911 (see Trove). Available at the State Library of SA.*

Book: McDouall Stuart's companion
Wilson, Shirley (compiled by Annabel Price & and Jill Watt) *Billiatt: John William Billiatt: Explorer, Adventurer and Tutor*, Print Solutions SA, 2006. He was the youngest member of McDouall Stuart's sixth expedition.

Books: NT regions
Forrest, Peter & Sheila *In the Middle of Everywhere – A History of Elliot and District* Darwin, Shady Tree, 2011.
Jones, Peter *Jones Store, Newcastle Waters, Northern Territory: A Social History of an Outback Store*, McDowall, Queensland, 2016.
Parks & Wildlife Commission of the NT *Malakmalak and Matngala plants and animals – Flora from the Daly River area*. Northern Territory Botanical Bulletin 26, 2001.

Books: Gulf Country
Camfoo, Tex & Nelly *Love against the Law* – autobiographies of Tex and Nelly Camfoo, Aboriginal Studies Press, 2001.
Merlan, Francesca, and Dirngayg, Amy *Big River Country: Stories from Elsey Station*, Alice Springs, IAD Press, Northern Territory, 1996.

Article: Gulf Country
Historian Tony Roberts has written extensively about the Gulf country. Here is one article printed in The Monthly, Nov 2009, pp 42–51, 2009. 'The Brutal Truth: What Happened in the Gulf Country'.
https://www.themonthly.com.au/issue/2009/november/
1330478364/tony-roberts/brutal-truth#mtr

Website: Land Rights
Visit the Central Land Council website www.clc.org.au for an explanation of the Land Rights Act, legislation exclusive to the Northern Territory.

EPILOGUE

Book
Hancock, David *Adelaide to Darwin railway – A Vision Fulfilled*, Skyscans Australia and Sprout Creative, 2004.

MISCELLANEOUS

Aboriginal authors
Including: Tony Birch, Tara June Winch, Melissa Lukashenko, Anita Heiss, Alexis Wright

More authors
Any book by Kim Mahood. *A good start would be her latest, a collection of essays* Wandering with Intent, *Scribe, 2022, winner of the 2023 Age Book of the Year award for nonfiction.*
Watson, Don *Caledonia Australis*, Penguin, 2009. *This is about Scottish Highlanders on the frontier of Australia.*
Anderson, Thistle *Arcadian Adelaide*, reprinted by Wakefield Press, 1985. First

printed in 1905.

A couple of books by Australian journalist, travel writer and novelist Ernestine Hill (1899-1972): The Great Australian Loneliness – *reprinted, Royal Exchange, NSW ETT Imprint, 2021.* The Territory (1901).

Dysart – McDouall Stuart's birthplace

Swan, Jim and McNeill, Carol *Dysart – In days gone by*, The Dysart Trust, 2013.
Swan, Jim and McNeill, Carol *Dysart – A Royal Burgh*, The Dysart Trust, 1997.

PEOPLE AND PLACES

ACKNOWLEDGEMENTS

This book evolved from a solo road trip through the centre of Australia. From the party to wave me off on my adventure back in 2015 to publication, I've gathered a grand list of friends and supporters who have helped to make *McDouall Stuart hitches a ride* happen.

I have to start with a big thank you to my family in Australia who encouraged me to do the trip in the first place. And my family and friends back in Scotland, who knew that any visits would lead to a library at some point. Thank you all for cheering from the sidelines, always.

I've had lots of mentors and highly recommend the concept of asking for help. Anne Skipper has been on my case from the start and is still coming up with clever ideas to keep the adventure alive. Heather Nimmo and Anne-Marie Smith took me and my book on a couple of years ago. Those fortnightly meetings in the State Library café, to talk about plot and dialogue and other writerly issues, kept me sane, focused and happy to get back to the task. Friends for life. Unofficial mentors can be found lurking in the various groups that have helped me keep a good work-life balance over the years, sharing common pleasures of swimming, cycling, walking, yoga, reading and writing. Thank you for the distractions and the genuine 'where are you at now' encouragements.

Beta readers, editors, proofreaders and fact-checkers were thankfully into multi-tasking. So many have been with me for the long haul, including Renate Nisi and David Meldrum, who read early drafts, Sandra Dann who read one of the last ones, and Linley Joomjaroen who has spent hours with me, pouring over both early and later versions. A special shout-out to friend and fellow writer Emma McEwin who did so much more than my request to 'cast her eye' over an early draft with excellent structural advice; Rick Moore, President of the John McDouall Stuart Society who read at least two drafts and saved me from a couple of blunders; and Colin Harris, a past President of the Royal Geographical Society of SA and President of Friends of Mound Springs volunteer group, who helped me differentiate my grasses from my sedges and pointed out the over-use of the word 'spurt' when writing about the mound springs.

It was lovely to reconnect with people I'd interviewed on my trip from Adelaide to Point Stuart, who were happy to check their sections, particularly Lyn and Gordon Litchfield, Pete (the postie) Rowe and Heather Rose Richardson at the Stuart's Tree Fishing Camp. Thanks also to the numerous people connected to various Aboriginal entities including Ross Williams in Tennant Creek; Max and Mabs Gorringe, former managers of Elsey Station; and, in particular, three people connected to the Arabana Aboriginal Corporation, Elder Sydney Strangways, Dr Veronica Arbon and Aaron Stuart, a director of the Corporation, who went out of their way to show interest in my project.

Those at the pointy end of self-publishing have been indispensable. Special mentions go to Amanda Ryder, friend first, my copy editor second, who rescued me from a serious case of hyphenitis and other dilemmas; Jan Finlayson, whose maps within the book are exquisite and tell stories all on their own; and Arjan van Woensel, who 'got' me from the first phone chat and came up with a cover I love and an internal layout that gives the book that final polish. You all lifted the look of this book beyond my expectations.

On a practical level, I also want to thank Rob Davey, owner of Complete Ute and Van Hire in Adelaide, who introduced me to Bad-Ass Betsie and made sure I was prepared for the trip, and the remote nursing organisation, CRANAplus, which I write for. They helped me organise interviews with health workers along the way, opening doors to meet locals, and gave me work at their conference in Alice Springs, mid-trip, which helped to fund it.

I must thank all the people working in libraries and museums in Adelaide, Alice Springs and Darwin who gave me their time and knowledge to help me locate resource material. What would writers do without them?

And, more recently, Bibi Medved used her graphic design and marketing skills to help me reinstate my blog *Along those lines* https://alongthoselines15.wordpress.com/and created my beautiful Stuart's Pea logo; and Snap Printing in Flinders Street, Adelaide, where I got help with digitising the maps.

This has been a rollicking journey that I will treasure forever. I thoroughly encourage everyone who has an adventure in them, whether it's a physical or a creative project, to get started.

FINAL NOTE: I've done my utmost to check details and any mistakes in this book are entirely mine. I've also grappled with words I've used in the book from primary sources that would not be condoned today. I left them in, as otherwise I would be joining the ranks of the whitewashers. I also take full responsibility for the final choices for the spelling of some place names.

ABOUT THE AUTHOR

Rosemary Cadden has worked as a journalist, media adviser and PR consultant with a special interest in environmental issues and history.

Since arriving in Australia in 1975, history has been a constant thread in her work and projects. She wrote and narrated a weekly segment for Adelaide University radio (then 5UV) called '100 Years Ago in South Australia' and was a History Writer for the Advertiser in the 80s.

In 1994 she worked for the Women's Suffrage Centenary Secretariat; much of that work is available on the State Library of SA website, including the article 'The Three Waves of Feminism'.

In 2009 she was commissioned to write *Building South Australia* to mark 125 years of the Master Builders Association in the state. Chapters include the 1920s story of the Colonel Light Gardens suburb, which is now a State Heritage area, and the post-war heyday of the SA Housing Trust. It was when Rosemary worked for Aboriginal organisations in South Australia and the Northern Territory in the 1990s that she was drawn to explore the effects of settler colonisation.

Rosemary is a member of the Australian Society of Authors, the Alliance of Independent Authors and WritersSA. Her cookbook *Making a Meal of It: smart ways to buy, store and use up food*, co-written with Jane Willcox, is published by Wakefield Press. During South Australia's 2023 History Festival, her sold-out talk, 'Fake news is as old as history' presented examples of misinformation, downright lies and long-lasting myths that she questions in this book.

For more information: rosemarycadden@gmail.com
Blog: Along those lines: https://alongthoselines15.wordpress.com/
Facebook: https://www.facebook.com/rosemary.cadden